Nerve Endings

ALSO BY RICHARD RAPPORT, M.D.

Physician: The Life of Paul Beeson

NERVE ENDINGS

The Discovery of the Synapse

Richard Rapport, M.D.

W. W. NORTON & COMPANY
New York · London

FRONTISPIECE. Cajal in his laboratory in Valencia.
From Lain Entralgo, "Ramón y Cajal 1852–1934."

Hayden Carruth, excerpt from "Essay on Death" from *Collected Shorter Poems 1946–1991.* Copyright © 1991, 1992 by Hayden Carruth. Reprinted with the permission of Copper Canyon Press, P.O. Box 271, Port Townsend, WA 98368-0271.

"Drawing of motoneuronal perikeryon" from Raphael Poritsky, "Two and Three Dimensional Ultrastructure of Boutons and Glial Cells on the Surface of the Motoneuronal Surface in the Cat Spinal Cord," *Journal of Comparative Neurology* 135: 423–52. Copyright © 1969 Raphael Poritsky. Reprinted by permission of Wiley-Liss, Inc., a subsidiary of John Wiley & Sons, Inc.

"Functions of the Lobes of the Brain" diagram from *A Primer of Brain Tumors*, 8th edition, © 2004. Reproduced with the permission of the American Brain Tumor Association.

For information about permission to reproduce selections from this book, write to Permissions, W. W. Norton & Company, Inc.,
500 Fifth Avenue, New York, NY 10110

Manufacturing by Courier Westford
Book design by Charlotte Staub
Production manager: Anna Oler

Library of Congress Cataloging-in-Publication Data

Rapport, Richard L.
Nerve endings : the discovery of the synapse / Richard Rapport.— 1st ed.
p. cm.
Includes bibliographical references and index.
ISBN 0-393-06019-5 (hardcover)
1. Synapses—History. 2. Ramón y Cajal, Santiago, 1852–1934. 3. Golgi, Camillo, 1843–1926. I. Title.
QP364.R37 2005
612.8—dc22

2005000942

W. W. Norton & Company, Inc., 500 Fifth Avenue, New York, N.Y. 10110
www.wwnorton.com

W. W. Norton & Company Ltd., Castle House, 75/76 Wells Street, London W1T 3QT

1 2 3 4 5 6 7 8 9 0

In Memory of

MARGARET SHEA GILBERT,
> *who first showed me the nervous system*

AND

ELIZABETH CROSBY,
> *who showed me its mysterious beauty*

Brains, bones, blood, synapses, little electrical currents—
but where do the souls go?

—HAYDEN CARRUTH

It is something to observe, but it is not enough. . . .
As I have repeatedly shown, observation itself is often a
snare: We interpret its data according to the exigencies
of our theories.

—J. HENRI FABRE

Contents

Preface

THE DISCOVERY of the synapse, that tender junction where nerve cells communicate with each other—but don't quite touch—forms the basis for modern neurobiology. From the moment in the early nineteenth century when it became just possible to see the cells of the nervous system through improved microscopes until a majority of histologists (anatomists who study cells and tissue) finally accepted the theory of the neuron and synapse a hundred years later, an enormous amount of arduous, intellectually magnificent work was done by scientists often very unsure of what they were trying to find. Between about 1830 and 1890, the technology and the body of evidence grew so rapidly that the accepted opinion about the microscopic anatomy of the nervous system had been completely rewritten by the turn of that century. The discovery of the synapse, the gap between nerve cells, revolutionized not only basic biology and the specific disciplines of neuroanatomy and physiology, but also clinical medicine and psychology; reverberations even reached philosophy. While dozens of great scientists contributed to the views that made this discovery possible, one unknown Spaniard, working alone, identified the synapse itself.

Some of the early terminology used to describe what we now identify as neurons (nerve cells), axons (a single fiber conducting away from the cell body), dendrites (multiple fibers on a cell body that receive information from the axons of other cells), and synapses is confusing. Indeed, because most of the current terms used to describe a nerve cell and its parts were either being made up or altered as the neuron theory was becoming accepted, I have defined these various usages in a glossary at the end of the book. Most important, the synaptic space itself—bounded by the terminal membrane of the axon (presynaptic) and a similar membrane across the space on the receiving dendrites (postsynaptic)—could not be identified until the era of electron microscopy. Therefore, when nineteenth-century anatomists sometimes referred to axons *contacting* cell bodies or dendrites, they used the word to describe single neurons stimulating other neurons by an electrical impulse generated down a fiber, sparking across a gap, and exciting receptors on the next cell. For them, "to contact" actually meant to be in touch with, rather than to touch. It took until the middle of the twentieth century for the chemical nature of synaptic transmission to emerge out of laboratories by then rich with electron microscopes, microelectrodes, and modern biochemistry.

In the summer of 1964, in the ground-floor laboratory of one of the University of Chicago's Gothic science buildings, I first saw a neuron myself. That is, I saw its corpse, stained with silver and laid out like a branching tree felled onto a glass slide. While I had studied drawings and photographs of these cells, with their long, thin axons wandering across the page in search of the stubbier dendrites sprouted from the surface of neighboring neurons, I had never before focused down the eyepieces of a high-resolution microscope and actually looked directly at this fundamental unit of the organ that makes a being human. For a twenty-one-year-old college student this sight was startling and beautiful, its perfect smallness creating the same kind of reverence one feels for the immensity of the universe.

That summer, preparing to write an undergraduate honors thesis on the fundamentals of neural connection, I had managed to land a fellowship in the lab of a neurologist who was also an experimental neuroanatomist. First he taught me the technique for making stereotaxic lesions—that is, killing a tiny area of an animal's brain by passing a small current through a fine wire electrode surgically inserted into a specific cluster of neurons. I also had to learn the methods for staining brain tissue with silver. Finally, I practiced how to examine the slides I had made microscopically, just as the early histologists had done. These tools would enable us to map the course of the resulting degenerated neurons, thus showing where the axons from those killed cells came to an end by following the trail of silver. That neurons ended at a synapse was, by then, a universally accepted fact in neurobiology. Our laboratory was investigating the chemical means by which these cells spoke to each other across the gap.

Making the lesion required first immobilizing an anesthetized rat very precisely in a head holder. I made the fine wire electrodes myself and insulated them with lacquer, except at the tips. Detailed anatomical manuals were available that permitted us to position an electrode carrier in three dimensions, and to locate a specific cluster of cells deep in the animal's brain. I calculated the coordinates, drilled a tiny hole in the skull, and inserted the electrode to a precise depth below the surface of the outer layer of the brain called the cortex. Much to my surprise, I had discovered a use for trigonometry. Finally, I passed a current through the electrode for a few seconds, creating the lesion. When a collection of neuronal cell bodies (called a nucleus) is killed in this way, their axons also die, and can be stained by one of several techniques using solutions of silver salts.

These staining methods, which I found enormously difficult to master, were derived from older techniques known by the exotic names of their discoverers: Golgi and Cajal. (Throughout the manuscript, I have referred to Santiago Ramón y Cajal by the usual American abbreviation of his last name. Other Spanish surnames I have kept in the more proper form.)

When I initially gazed at my creations down the eyepieces of a flawless two-headed, binocular Zeiss microscope, I wasn't sure what I was seeing. By the end of the summer, though, I could find my way around the slides and knew enough basic anatomy of the brain, stereotaxic surgery on rodent patients, and silver-staining techniques to design the experiments that were to occupy me for the next year. I could recognize neurons and was experienced enough by then to distinguish them from neuroglial cells. Glia are supporting cells that do not electrically conduct but provide important services, including insulation, repair, and protection, to the cells that do. The dendrites that collect messages into the neuron's cell body were visible to me, and I could find the long, graceful single axon departing for the next cell. I already grasped the electrical depolarization by which neurons conduct information in one direction, dendrite to axon, and the basic mechanism for the transmission of information between cells. Understanding the exact biochemistry of these synaptic transmitters, however, had been the object of our research.

Twelve years later, I had completed medical school, a neurosurgical residency, and a fellowship at the National Institutes of Health. Finally I had a little time to spend in used-book stores. One day on a dusty, seldom-visited shelf, I found a copy of Santiago Ramón y Cajal's autobiography, *Recollections of My Life*. After reading the first paragraph, I couldn't stop. In all those years of preparation, I had never anticipated this voice reaching across more than a hundred years and telling a story of passion, discovery, argument, and, finally, acceptance of a new truth. Arrested by the simple beauty of the prose, I stood captured in the aisle for an hour. The writing was so graceful and vivid that I quickly entered into the life of an ungovernable boy growing up in the hills of northern Spain, and watched as he matured on the page into a rare artistic and scientific genius.

Cajal later not only described much of the microscopic anatomy of the nervous system, but also proved that axons halt just short of their destination and communicate with the next cell across—a gap. The existence of this gap (later named "synapse" by the English physiologist Charles Sherrington) had for years been the subject of rancorous debate. The quarrel climaxed during Cajal's lifetime, and was all but settled by the time he died. When I finished the last page of his elegant autobiography that week, I had an epiphany. It struck me after I read this passage:

> For all those who are fascinated by the bewitchment of the infinitely small, there wait in the bosom of the living being millions of palpitating cells which, for the surrender of their secret, and with it the halo of fame, demand only a clear and persistent intelligence to contemplate, admire, and understand them.

That enchanted description of the investigator's soul convinced me that Cajal was a person born to spend his life devoted to a study of basic neurobiology, and that I was not. I became a clinical neurosurgeon instead, and realized the joy of knowing the neuron in a different way.

Writing for a general audience about this breakthrough, I have tried to minimize the technical jargon, lists of scientists, and the academic use of complete citations in footnotes. Though I have kept to the historical context, this is not a detailed history or a complete biography of all the researchers whose labors revealed the neuron and the synapse. I have made every effort to be fair to the memory of all those who participated in what became an acrimonious scientific controversy.

RICHARD RAPPORT
Seattle
September 2004

Acknowledgments

Robert Yates Moore, my mentor at the University of Chicago, taught me silver stains and introduced me to the neuron, the scientific method, and my wife. I owe him a lot.

I first met the great neurobiologist Arnold Scheibel in 1970, when I was a fellow at NIH and he was at the UCLA Brain Research Institute. A coincidence brought us together again thirty years later. Knowing nothing of my long interest in the man, the shelf of books, or files of notes about him, Dr. Scheibel stepped back and exclaimed, "You look like Santiago Ramón y Cajal!" To the extent that many middle-aged, thin, bald men with gray beards resemble one another, he was right. That meeting and our long discussion about Cajal, Golgi, and silver stains encouraged me to complete this project, a book I had long wanted to write. Dr. Scheibel read the manuscript in two different early forms and corrected my most glaring errors. I am grateful to him for all his kindness, patience, and profound knowledge of the subject.

My colleague John Howe, a neurosurgeon and neurophysiologist, also read an early draft, and his criticism improved the writing. I am thankful to him for that, and the forty years of his friendship. In a final effort to expose any scientific errors I might have made, another old friend, University of Washington neuroanatomist Les Westrum, read a late draft of the manuscript. If any mistakes escaped these readers, the fault is mine.

Jessica Papin, my agent at Dystel and Goderich, found me the right publisher and made the entire project much easier. Amy Cherry, my fine editor at Norton, gently improved both my language and style. Both of them made this a better book.

Nerve Endings

ONE

A New Way
to See

A SMELL OF THE EARLY-SPRING morning floated up from the damp pavement of Barcelona's Las Ramblas. Old men watched as bird sellers carefully hung arrangements of cages in front of their shops and flower vendors opened the shutters of their stalls for business. As he strode along the wide, tree-lined boulevard, Santiago Ramón y Cajal would have seen the top of the new two-hundred-foot high monument to Columbus just erected in the Puerta de la Paz, and beyond it the sea. A slim, compact man, Cajal was already balding, and his close-cropped beard framed a chiseled face dominated by deeply set, intense eyes.

Along the way, he may have stopped to admire the elegant architecture of the buildings that lined the broad avenue, or even to sketch some of the Catalonian faces he saw there. The noted local professor was an accomplished artist, and he understood the world with an artist's eye. Having been born of peasant parents in rural Spain himself, he would have appreciated the vigorous charm of the vendors and other early risers eating and debating around the sidewalk tables as the day began.

The International Exposition of 1888 was under way. The recently appointed professor of anatomy at the University of Barcelona would likely have agreed to meet his chemistry professor friend, Victorino García de la Cruz, at the pavilion where George Eastman's new Kodak camera was on display. Cajal was an enthusiastic amateur photographer anxious to learn for himself if this one-button device really took photographs on small strips of film that rolled through the camera, rather that on emulsion plates. That morning the pair would also have been able to hear the talk by a physics professor about Heinrich Hertz's discovery that electromagnetic waves exist in space around a discharging Leyden jar, a sort of primitive battery. Hertz had just proven that these waves are propagated at the speed of light, just as James Maxwell had predicted fifteen years earlier.

Cajal and his companion would have enjoyed the entire morning at the exposition. The anatomist had to spend an hour or two at the exhibit of his own microscopic preparations, for which he was to be awarded the exposition's gold medal. With genuine animation, he would have delighted in explaining to onlookers the beauty and power of the recently perfected silver-staining techniques that displayed brain cells with new clarity. The rest of the time the two academics would have spent wondering together at the modern discoveries in biology, chemistry, and physics that surrounded them. Some of these ideas were by then pushing aside venerable classical theories.

When they met to eat with a larger group of professors, writers, and politicians, the two men regularly debated such controversies. These intellectual friends gathered to spend a few hours in the middle of the day at the Café de Pelayo, a restaurant near the Hospital of Santa Cruz, where the Faculty of Medicine was located. The meal included not only the choice of dozens of *tapas* but also debate just as varied about new theories of electromagnetism, evolution, photography, histology—and, of course, politics. The reign of King Alfonzo XII, followed by the regency of his widow, had provided a momentary stability to the frequently wild swings in Spanish political life. But the conversation this day concentrated on new theories in physics that seemed to challenge the authority of Newton, who had unhesitatingly requisitioned divine intervention to solve his equations, sometimes complicated by too many zeros and infinities.

After they ate, Cajal and his colleagues played chess and talked for another hour before they all returned to work. For the scientist, this was his time to sit alone at the microscope, before students arrived to join him at home in his laboratory, adjacent to the garden where he kept a few experimental animals. In order to supplement his small university salary by a few dollars a week, Professor Cajal held late-evening private lessons with several capable students especially interested in the study of cells and tissue—the relatively new science of histology.

One of the microscopists who came to study that night, a young Dr. Pi y Gilbert, was particularly anxious to look at the slides of mouse cerebellum his teacher had been preparing over the previous weeks. The process of fixing, cutting, and staining brain tissue was a tedious one. The innovative silver-

staining method perfected by Cajal, his student realized, outlined brain cells with a new clarity and permitted an examination of not only the body of these cells but also their elusive
fibers. Pi y Gilbert had read in his textbooks that the basic silver stain had been discovered by the famed Italian anatomist
Camillo Golgi at the University of Pavia, five hundred miles to
the northeast. While the distance between northern Italy and
Barcelona was great geographically, it was even farther intellectually at the end of the nineteenth century. Over the year
since Cajal had been named Professor of Normal and Abnormal Histology, most of the students and faculty at the medical
school had come to admire him and to respect his work, even
though most of them were incapable of understanding his
novel ideas about the construction of the nervous system.
After all, the scant science being done in Spain could not com-
pare to the brilliant work produced at the much more prestigious universities of Germany, Italy, France, and England.

The garden door stood ajar as Cajal adjusted the mirror
and focused light onto the stage of his prized Zeiss microscope. This instrument, the best available, had been a gift
from the provincial government of Zaragoza in recognition of
his work in quelling the local cholera epidemic of 1885, the
same year the German Robert Koch identified the microbe
causing that disease. In preparation for the arrival of the students, his wife gathered their children into the back of the
house and sent the younger ones to bed. Señora Cajal had
also cleared the kitchen of food and dishes so that her husband and his students could heat paraffin in a pan over the
stove if they wanted to make new slides.

A few weeks before, Pi y Gilbert's friend and fellow histology student Durán y Ventosa had found a paper recently published by Cajal, and they both were eager to discuss it with him. The brain and spinal cord are too complex, the students had complained to each other. Many of the theories about the structure and function of the nervous system seemed to them contradictory. Durán y Ventosa agreed with his companion that Cajal's new silver-staining method was splendid for showing brain cells, but neither could understand where that led him. They knew he claimed that the fibers of these cells ended, but if they did, how could information be transmitted at such a dead end? In any case, why would Cajal propose to argue with the Italians and Germans whose science is so far beyond our own? they wondered.

The Opera House was filling with theatergoers as the two young men made their way across Las Ramblas and down the Gran Via Diagonal. When they arrived at the home of their teacher, he was already gazing through his microscope. Beside him on the table were an ink pot, good paper, and a fine-nibbed pen. As his several pupils entered through the garden, Cajal finished sketching the ghostly outlines of brain cells he had been staring at all afternoon. Their study began immediately.

Cajal taught by demonstration as he moved the slide, first targeting the large body of a cerebellar Purkinje cell. These brain cells (neurons), with thousands of axonal connections to their graceful arbor of dendrites, had been found by Jan Purkinje in 1837. Because they are relatively large, such cells made themselves available for study even in unstained sections contemplated through simple microscopes. Cajal had already introduced his students to Purkinje cells in their regular anatomy classes, and they were therefore prepared to move past the basics. Next he showed them an axon leaving the cell body of one of these neurons, pointing out the unique, large fiber outlined with silver at one end of the cell. On the opposite side he found the feathery dendritic processes, neatly arranged almost flat in one plane, that received messages from other neurons.

At this point in the lesson, Pi y Gilbert and his colleagues would have been skeptical, especially about the implied direction of electrical flow through the cells. The great Golgi himself had proven, they believed, that the cells of the nervous system are a network (or reticulum), and if they are all connected one to another, the current of information must flow in all directions. How else could messages be sent and received?

This was the subject of Cajal's paper, read intently by the two students, however much it had bewildered them. In fact, the review of histology containing this article was Cajal's own journal, published in Spanish and locally distributed, but read almost nowhere else in Europe. The students had the confidence of received wisdom to stiffen them when challenging their teacher on this subject. Although Cajal was admired in Barcelona, his theories were hardly accepted widely even in Spain. They were also well aware that most of the important anatomists devoted to neurohistology, including Golgi and the acclaimed German Albert von Kölliker, believed that all the cells of the nervous system were directly connected to one another. Neural transmission seemed to require this sort of cellular hand-holding, and the reticular theory was a popular explanation for how information moved throughout the nervous system.

Excited by the opportunity to show off, the poised Durán y Ventosa (son of an ex–minister of state) and the brilliant Pi y Gilbert may have speculated aloud about where their teacher could see what he called his "contacts." Moreover, how could the fibers rising from the brain stem into the cerebellum find their way to the Purkinje-cell bodies? After they arrived there, what happened to them? How might they announce them-selves to their neighbors? Surely everything must be con-nected to everything else; otherwise there would be barriers, and information would simply stop.

Looking up, Cajal must have smiled slightly and pushed his chair away from the table strewn with slides, blocks of paraffin, a pen, and pages of drawings. He enjoyed this debate with the younger men and welcomed their enthusias-tic interest in a subject to which he had dedicated his life. "Instead of elaborating on accepted principles, let us simply point out that for the last hundred years the natural sciences have abandoned completely the Aristotelian principles of intuition, inspiration, and dogmatism," he advised them firmly but not unkindly. With a slow patience, he described the places he had observed nerve fibers ending *near but not continuous with* the next cell body. In this way he explained his belief that a fundamental cellular unit defined the neuron theory. The axon ends, and therefore the nervous system is not all directly joined in a reticulum. Information from other cells enters neurons through dendritic tentacles on their cell bodies and exits via a gap at the end of the long axons that leave them. Thus the impulses travel in one direction in cells of the brain and spinal cord, as well as in peripheral nerves. He slid his chair forward to the table again and pointed out exactly where he saw this cellular gap.

Even though Pi y Gilbert and the others were never absolutely certain that they could see the same things that Cajal assured them the slides clearly revealed, they grew to believe that their teacher saw them. And they came to understand that Cajal was observing and describing the nervous system in an entirely new way. Although almost no one realized it, this was a new age of Spanish discovery.

HUMAN BEINGS HAVE BEEN investigating their own brains at least since they began to speculate. Exploration of the way our minds work, and the way they work our bodies, has traveled from magic and superstition to philosophy and then, with a hiss from the viewing chamber of the electron microscope, to subcellular anatomy and biophysics.

The microscopic neuron, or nerve cell, is the basic unit of our nervous system, that internal communications network evolved both to get the news from the world and to broadcast it to muscles and organs. When it was initially glimpsed through light microscopes early in the nineteenth century, scientists began first to recognize and later to name this cell, at once ending thousands of years of contradictory philosophical conjecture and creating new disputes. This was a moment in the sciences when the absolute world of Aristotle and Newton began to spin uncertainly, and new technology permitted quantifiable observations not only in biology but in chemistry and physics as well. Modern science was being born. In the last half of the nineteenth century, this world of observation was dominated by confident German academics, most of whom would have enthusiastically agreed with Prince Otto von Bismarck when he told the Reichstag on February 6, 1888, "We Germans fear God, and nothing else on earth."

Plato, who had only a few tools at his disposal, and almost no data, constructed a cosmology built on the four known Greek elements and on various shapes. As he conceived the fundamental units of this universe to be geometrical figures, he taught his students that the body was fabricated mathematically. The Greeks struggled to locate the resin binding soul to body, a problem that has always occupied philosophers and theologians. Plato believed it to be the purest form of the triangle that is the elemental component of earth, air, fire, and water, and that this substance resides in the "marrow" of the cranium and spinal canal—what we now call the central nervous system. This marrow of perfect triangles was for Plato the essence of matter. Therefore, he reasoned, it glued the soul and the body together. As a metaphor, the image of a Platonic nervous system was an accurate description: It does more or less hold things together without necessarily marrying them. The batterylike switching on and off of neurons—although they aren't triangles—accounts for all our mundane mechanical bodily functions and all our yearnings.

For the next two thousand years, philosophers continued the struggle to imagine their own brains. Prior to the development of useful microscopes in the late seventeenth century, gross anatomy supplied the only real data that could be collected for investigating this mystery. Dissection and pathological studies had already led to an approximate understanding of how the large organ systems functioned. The cutting open of human and animal bodies did not reveal, however, that neurons even existed or that the nervous system was electrical. Whereas the heart looks like a pump, and the collecting system in a raw kidney suggests a filter, what does the soft, grayish lump of trembling brain with a fluid-filled center chamber propose?

The general relationships between form and function were by then established, so struggling to deduce a purpose by studying the squishy contents of the cranium and spinal canal took fanciful directions. Descartes, for example, who wrote the first European text on physiology, was still really a Galenist (in that he interpreted bodily functions in terms of theory more than observation). His dissections led him to conclude that the centrally located but diminutive pineal body had a sort of curator function, determining which thoughts and sensations would be allowed to ascend into higher consciousness. Conveniently situated by God in the center of the brain, the pineal directed this flow, he speculated, by moving back and forth to control the passage of sensations, passions, and philosophies. Movement by this gland in sentient beings, a duty that Descartes equated with the soul, intermittently blocked the various openings of the ventricular system—that lake inside the substance of brain where spinal fluid is made. In this scheme, certain ideas and feelings were thereby directed to particular anatomical places by the pineal. Perhaps it was in part this peculiar lesson of Descartes' that helped overexcite his last pupil, twenty-two-year-old Queen Christina of Sweden, and propel her on to a series of follies. A passionate student with a dynasty-halting aversion to marriage, Christina converted to Catholicism and abdicated her throne, thereafter living lavishly, if not always wisely, in Rome.

At the same time he had the pineal directing traffic in the brain, Descartes believed transmission in nerves occurred when fine particles moved inside them, traveling through what he conceived to be an aqueduct constructed of hollow tubes. All the theories about neural function—from Pythagoras to Descartes—concluded in philosophical speculation about the *look* of the brain, spinal cord, and nerves that was limited by scale, and a view of the world based on ideas rather than observation.

Microscopes changed everything. The brain was suddenly no longer simply an ineffable pudding. Instead of struggling to determine where motor and sensory functions were stored, not to mention anger, love, language, and theology, guesses about the function of this half-pound occupant of the skull gave way to looking at its basic units.

In attempting to understand this organ and its newly dis-

Functions of the lobes of the brain

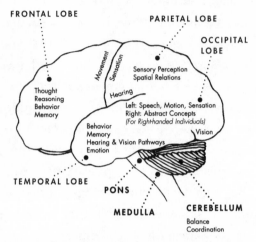

Schematic drawing of the left hemisphere of the human brain, showing approximate areas of certain functions. *From American Brain Tumor Association,* A Primer of Brain Tumors.

covered cells, the "hard wiring" of the brain, spinal cord, and nerves, humans applied their own intelligence to the study of themselves. This strategy, it has more than once been suggested, may have limits. In the seventeenth century, Pascal thought, "Man is to himself the most wonderful object in nature; for he cannot conceive what the body is, still less what the mind is, and least of all how a body should be united to a mind. This is the consummation of his difficulties, and yet it is his very being." Even so, beginning in about 1850, during one of those convergences of technology, tenacity, genius, and luck, histologists and physiologists began to see actual neurons and to understand their electrical nature. Rather than simply imagining what they might be, it became possible to describe their microscopic anatomy and guess at their function.

Following the invention of the first microscopes, science pushed aside metaphysics, and technical developments allowed researchers to see smaller and smaller bits of nervous tissue in greater and greater detail. Galileo probably made not only the first telescopes (attracting so much unwanted attention by the Inquisitors in Rome) but also their converse—simple microscopes. The latter, in part because they were much less satisfactory instruments than his telescopes, were a lot safer for him. Over the next two hundred years, these first microscopes suffered from hazy distortions of the image (spherical aberration) and the separation of light into colors (chromatic aberration). These flaws inhibited accurate description, therefore, and failed to launch histology as a useful branch of science. Better optics improved the instruments, but still tissues were examined fresh and unstained, techniques that severely limited their detail.

The systematic observations made by the Dutch lens maker Antony van Leeuwenhoek in the late seventeenth century, including his descriptions of an optic nerve from a cow, opened the window on what he called "wondrous small creatures." About the same time in Italy, Marcello Malpighi had a microscope good enough to examine embryos, including their rudimentary brains. At first this was exhausting labor-intensive work, requiring meticulous preparation of the slides and then hours of squinting observation. But by late in the eighteenth century, perfection of compound (two lenses), achromatic microscopes permitted close examination of still-unstained sections of brain and of single cells teased out by hand, separated from their fellows with a needle.

The change in scale and the 1858 introduction by Rudolf Berlin of hardening brain tissue before it was cut, and then staining it, resulted in new, unforeseen complexities introduced by the techniques. Early simple staining methods colored so many cells and fibers that the individuals got lost in vast crowds, and the maze could not be followed. While the cells revealed themselves, their processes did not. By using better techniques that stained some (but not all) individual cells on a slide, suddenly scientists could formulate detailed, if often comically imprecise, descriptions of central-nervous-system architecture, some as fanciful as Descartes' pineal prelate.

About the same time, Luigi Galvani hooked up two electrical wires to the spinal cord of a decapitated frog and, by passing a current through them, made the animal's legs jerk. He concluded that nerves conduct electrically, in that moment creating the discipline of neurophysiology.

But knowing that nerves cause muscle to contract by some electrical mechanism and seeing parts of a few stained neurons in two dimensions on a glass slide is hardly understanding the complex, three-dimensional anatomy and physiology of billions of these cells. It is further still from knowing how they are organized, much less how they communicate with each other or enable the organism to communicate.

Finding the anatomical basis for this intimacy between neurons—finding the synapse—was a search conducted by hunters who weren't really certain of their quarry, using equipment just barely equal to the task. The arguments between these anatomists were among the most stimulating and bitter scientific battles of the late nineteenth century. Although there were dozens of great names involved in solving this mystery, the struggle between the reticularists (who believed that all the cells of the nervous system were *directly* connected) and the neuronists (who argued just as strongly that *gaps* existed between cells) is personified in the lives of the two leading histologists of the age, Camillo Golgi and Santiago Ramón y Cajal. Both of these men made great contributions to an understanding of the nervous system. And both saw neurons. One of them knew what they really were.

TWO

The Crazy Navarran

Beyond the cobbled streets of the village itself, there are two notable facts about Petilla de Aragon, which in 1852 belonged politically to Navarre on the southern slope of the Pyrenees. First, by one of those odd and passionately maintained geopolitical accidents particular to Spain, the village was physically located exactly in the middle of the province of Zaragoza, part of Aragon, not the Kingdom of Navarre. Second, the district doctor working in Petilla was Cajal's father. In reality, though, Justo Ramón Casasus wasn't exactly a doctor then, but rather a second-class surgeon, capable of performing minor operations and treating simple illnesses.

After his own parents died when he was about twelve, leaving their meager farm outside the village of Larrés to an older brother, Cajal's father was on his own. Resourceful and tough, he spent the next ten years apprenticed to a local barber surgeon who taught him to cut hair and let blood, the latter then still a popular treatment in the countryside for maladies from cough to cancer. Although some patients improved in spite of this treatment, Cajal's father had been able to study his master's books and knew that there was more to medicine than removing blood. At the age of twenty-two, he walked the two hundred miles to Barcelona, found work in another barber shop, and managed to attend courses at the university until he was partially trained as a physician. Today he would be a nurse-practitioner or a physician's assistant. After several years of this study, during which time he survived on a small wage and even smaller tips at the barber shop, Don Justo returned to the family home and began to treat patients. He was immediately successful and soon moved to the slightly larger town of Petilla. Two more years of saving enabled him to buy a small house and to marry his childhood sweetheart from Larrés. Although at the time of his first son, Santiago's, birth he was not a fully trained doc-

Petilla de Aragon where Cajal was born and lived until he was two. *From Cajal, Recollections of My Life.*

tor, he was a skilled and respected man, one of the few people with any education in the village.

Petilla itself was barren, austere, and remote, its sparse, modest buildings enclosed by thick stone walls originally built to stop Moors invading from the south and Goths from the north. This little town arose at the base of a huge mountain, and could only be reached by a hazardous, narrow path that followed a stream flowing out of the steep valley. Its fifty or sixty houses were humble enough, unadorned even by flowerpots in the windows and separated from one another by narrow, winding, and cluttered passageways. Decoration was an unaffordable luxury. The farmers of Petilla had painstakingly terraced from the mountainsides the only arable land, narrow plots that they continually had to defend from annihilation by the elements. Ruins of an ancient medieval keep dominated the hillside above the village; similar remains are still scattered in mountainous regions throughout Spain.

As if to match the rugged landscape, Cajal lamented that he and his three younger siblings, a brother, Pedro, and two sisters, Pabla and Jorja, had all inherited the plain looks and intensity of their father, not the smoky beauty and practicality of their mother. But he did not complain about the other character traits his father passed on to him. Nor was his father content to trust biology alone to make his offspring intelligent and ambitious. He made certain his son's genetics were supplemented by vigorous and demanding instruction, which he supervised himself.

Although he hadn't been able to afford a complete medical

education as a young man, Don Justo compensated with enthusiastic ambition, an iron will, and one of those remarkably indelible visual memories so useful in translating the drawings in textbooks into the blood, organs, and bones of surgery. He believed in hard work, diligence, and, because so much of his life took place in kitchens and bedrooms turned into temporary village operating rooms, in the enduring truth of gross anatomy.

Since he was the only doctor of any kind for mile after unpaved mountain mile, these characteristics sustained him as he labored at treating whatever ailments and injuries befell the peasants of his district. These illnesses and traumas were similar to those that Flaubert (the son of a physician) imagined his character—the limited public-health doctor, Charles Bovary—might be treating across the mountains in France at the same time. In fact, had Don Justo not been so repelled by the very idea of novels, he might have sympathetically read this account of a conversation between the pedantic and scheming apothecary Monsieur Homais and Madame Bovary's husband:

> Aside from the usual cases of enteritis, bronchitis, liver complaint, etc., our roster of illnesses includes an occasional intermittent fever at harvest time, but on the whole very little that's serious except for a good deal of scrofula, probably the result of the deplorable hygienic conditions in our countryside. Ah! You'll have to fight many a prejudice, Monsieur Bovary; every day your scientific efforts will be thwarted by the peasant's stubborn adherence to his old ways.

But even for the hearty Don Justo, who had arisen from that same peasant stock, the landscape around Petilla was bleak. Years later, following a visit to the place where he had been born, Cajal wrote, "Oh, the heroic rustics of our barren plateaux! Let us love them sincerely. They have performed the miracle of populating sterile regions from which the well-fed Frenchman or the rubicund and lymphatic German would have fled."

It was the will of the father, and what in time proved to be the genius of the son, that combined to produce rare qualities already apparent in the growing child. The family had moved from village to small town, increasing along the meandering route his calling took Don Justo. They settled for a time in Valpalmas of Zaragoza, where Cajal began school, although his father was his real teacher. In an effort to supplement lessons provided by the local teachers, Don Justo conducted extracurricular instruction in an abandoned shepherd's cave. There he sequestered himself with his son in order to avoid being interrupted by the sick for a few hours, as well as to concentrate, among the rocks, on the subject at hand. The lessons had a certain pragmatic rationale, and since France was but a few miles across the mountains, French was in the curriculum early. By the time he was six, Cajal not only could read and write his native tongue but also had mastered enough French for conversation, and could read and write the basic language of his neighboring country. Later he learned the fundamentals of geography, elementary physics, arithmetic, grammar, and classical literature. Whenever he read from The Odyssey, a story he had first heard there as a little boy, these Wordsworthian moments alone with his sturdy and pragmatic father could still transport Cajal, even as a very old man, back into the "recesses and windings" of that memory cave.

A difficult adolescence lurked ahead of this idyll with his father, along a halting evolution toward adulthood. The road was a perilous one for Cajal, and as painful a passage in the last half of the nineteenth century as it is for contemporary parents and their children. The beloved first son, an eager scholar and tractable child, was transformed suddenly into a sluggish dreamer, a novel reader, an artist, and—by the time Don Justo had finally completed his medical training and graduated as a physician to the bigger town of Ayerbe—a seeker of creative mischief.

While his father's degree and move to Ayerbe in the prosperous wine-growing country of Huesca promised advancement for the whole family, eight-year-old Cajal didn't initially thrive in his new community. It may have been his pure Castilian dialect that so contrasted with the local mixture of French, Catalan, Aragonese, and Castilian, or maybe the oddness of his clothes, which the locals looked upon as a "most disgusting affectation." The "little gentlemen" of the town did not admire Cajal's economical dress, without sandals or a handkerchief binding his head in accordance with the regional habits. Though his speech, clothes, and bashfulness were irritating to the children already established in the community, their attitude also brought Cajal a "certain morbid pleasure." He already knew he was different, if not intellectually aloof, and it didn't bother him unduly. Whatever branded him, his arrival was not welcome. Like many of us who retain those most emotional moments from childhood, Cajal remembered the first few weeks in his new town with painful clarity:

> My appearance in the public square of Ayerbe was hailed with general mockery by the boys. From mockery they passed to serious abuse. Whenever several of them were together and believed themselves certain of impunity, they insulted me, struck me with their fists, or tormented me by throwing stones at me. How brutal we boys of Ayerbe were! Why this idiotic aversion to the strange child?

The aversion and ill will softened, and though he may still have seemed strange to his playmates, Cajal was soon marauding through the streets with the rest of them, playing games, stealing fruit, and fighting with others in the manner of boys everywhere. Perhaps because of his oddness, combined with his own headstrong qualities and the ambition learned from his father, he quickly became their captain. "In the bottom of every juvenile head," he later wrote, "there is a perfect anarchist." Searching for escape from the rule of law, he also sometimes preferred to be alone. His explorations of the natural world around the edges of town taught Cajal to identify dozens of kinds of birds, and he could match at least twenty species with their specific type of nest. He earned his father's rare approval for one of these naturalist activities when he collected and classified examples of the eggs from thirty different kinds of birds. Unhappily, the heat of August soon caused them to rot in the compartments of their display box, creating a temporary sulphurous odor in the house and complaints from the whole family. He had better luck capturing the birds themselves, and studying them for a few days before letting them go.

Throughout his life, he considered this sort of play fundamental to development and one of the mechanisms by which children internalize and subdue the marvels of their world. Games, he later wrote, are an "absolutely essential preparation for life" and encourage vital maturation of the infantile brain. Cajal also believed that the element of danger added to this crucible acted as an enzyme, an idea now supported by psychologists and neurophysiologists studying child development and memory.

In spite of these quests, Don Justo's disappointment kept growing during 1860 and 1861. When his son abruptly converted from slings and mischief to a "madness over art," the transformation was hardly more satisfactory to Don Justo, who disapproved of art as much as he did of disorderliness. Cajal made the switch from petty theft and mayhem to painting by first drawing scenes of war and then graduating to images from the lives of the saints—preferring "the active to the contemplative ones." He most favored his namesake, Santiago (St. James), "the apostle, the patron of Spain and the terror of the Moor."

His father despaired and threatened. But as a practical matter, he decided to satisfy himself as to whether the boy had any artistic talent. There was only one art critic for Don Justo to consult in Ayerbe—a plasterer and house painter just arrived to repair the fire-damaged church. That reviewer,

A watercolor of the Hermitage of the Virgin of Casabas, near Ayerbe, painted by Cajal at the age of nine or ten. *From Cajal,* Recollections of My Life.

whom Cajal referred to as "the Aristarch" (a severe Greek critic of Homer), did not rate the boy's talents highly. This pronouncement had the effect of being uttered "*ex cathedra* by an Academy of Fine Arts." Paints, brushes, and colored pencils were confiscated, replaced by classical literature, geometry, and biology. In fact, though, the watercolors reproduced in *Recollections of My Life,* painted when Cajal was still a child, certainly reveal a gifted eye and a hand created for recording detail. What they do not disclose is the mind capable of seeing, understanding, and making clear what others could not imagine.

The negative judgment of the house painter did, however, end any further discussion of secondary education at art schools in Huesca or Zaragoza. Instead, Don Justo, arguing for economic security as a doctor, packed the boy off by horse-drawn wagon to live with his uncle in Jaca. He enrolled him there in the College of Esculapian Fathers to study for his baccalaureate (a course of study roughly analogous to American high school). In spite of the friars' supplemental application of the strap, the switch, or solitary confinement to their lesson plans, Cajal's desire to draw suppressed his appreciation for the value of Latin, geography, and mathematics.

Final proof for his father that the boy was hopeless came during the next school holiday. That summer of 1863, bored with constructing slings, bows and arrows, and other Stone Age implements of war, Cajal began to assemble a large cannon. This was not a toy but a real armament, much admired by his friends. Scavenging through the village, he had unearthed a large wooden beam, a discarded oilcan, wire, tarred cord, and a long, sturdy carpenter's auger. It was hard work to bore a straight hole down the length of the beam, then smooth it with sandpaper on a ramrod. For added strength, he wrapped the wood with cord and wire. Then in the hope that igniting the powder would not explode the entire enterprise, he fashioned a priming pan out of tin from the oilcan. Pleased with the result, Cajal and his accomplices decided not to mount the completed weapon on wheels as had been planned but instead—after "mature deliberation"—hoisted the creation onto an orchard wall and aimed it at the neighbor's brand-new garden gate across the lane.

Carefully loading their invention with gunpowder, cotton wadding, and cobblestones, they merrily blew apart the stout garden door. The loud report brought Cajal's neighbor dashing to the scene before the artillery brigade, entirely mystified that the cannon itself had not exploded, was able to withdraw. They were all apprehended, of course, and Cajal easily identified as the commander behind the assault. After consultation between his father and the mayor, "who had already complaints of other damage which I had done," Cajal was immediately jailed for three days. Incarcerated at the age of eleven, he was next tormented by the platoon of his unindicted co-conspirators and other villagers who came to mock him behind bars. This misery was not at all relieved by the jail cell itself, rich in "an overflowing flora and fauna," including fleas, bedbugs, lice, cockroaches, and a moldy pile of straw for a bed.

Even this unhappy experience was not the lesson it should have been, and Cajal's whimsy accelerated in new directions. To his father's horror the boy continued to neglect his study of Latin and Greek, favoring the more modern literature (Defoe, Cervantes, Dumas, Hugo, and Goethe) he discovered in the library of a neighbor. At about the same time, he was introduced to the new art of photography—a hobby that was to occupy his leisure for the rest of his life and have a great impact on his scientific discoveries. He had previously seen daguerreotype, a cumbersome process that produced a unique direct positive picture on a copper plate. However, the coaxing of a latent image from a mist of chemicals through the reducing agent's action on silver was a method that, he wrote, "positively stupefied me. The thing seemed simply absurd." This early photographic process allowed a negative image to be exposed more easily on a glass plate, and then repeatedly reproduced on paper. Taking the photograph appealed to Cajal's artistic nature, and the chemistry involved in developing and printing excited his inherent scientific curiosity.

But even this discovery did not remove him from the business of arms manufacture. "We were incorrigible," he wrote. Don Justo would have thought that judgment modest. He and his friends built yet another cannon from a large bronze pipe (this one did blow up) and then branched out into muskets. Appropriating his father's antique flintlock for a hunting expedition and conscripting his younger brother, Pedro, into the venture, Cajal had to figure out how to make powder, bullets, and birdshot for the relic before going to hunt rabbits and birds. A chemist after all, he bought sulfur, found some saltpeter in the cellar, and made charcoal out of charred soft wood. The shot he forged from discarded pieces of lead. Thus armed, the two boys climbed over the orchard wall, comically lugging the huge musket, and went in search of game. However, aiming the unwieldy weapon, lighting the fuse, and igniting the often damp powder invariably consumed so much time that their prey had safely vanished when the gun finally made its loud and gratifying report. Nothing engaged Cajal just then as much as his skill with this amateur weaponry.

Cajal's bad reputation spread throughout the village, so that girls leaving the schoolhouse were advised to avoid meeting him. The relatives of one of these little girls, a "timorous child," with "great sea-green eyes . . . and huge braids the colour of honey," repeatedly had their siestas interrupted by the son of the physician of Ayerbe. Not surprisingly, they vilified him in front of the girl. She ran away terrified and hid whenever she met the doctor's son, known by then as "the crazy Navarran." But perhaps too she admired him a little, this dangerous newcomer. And from the detail of the description in his autobiography, Cajal even then must have also noticed his future wife.

By the middle of the next term at Jaca, when Cajal was twelve, his father could no longer ignore the obvious. The boy was unlikely to become a classicist. He transferred to the Institute of Huesca where the course of instruction was broader, and less likely to be enforced with a cudgel by the fathers. And in addition to the usual courses, he could study art. Things improved. At the end of the term, he was able to go home having avoided trouble and succeeded in his classes.

When he returned to Huesca the following fall, his younger brother went with him. Don Justo, who now located his hope for producing a classics scholar in the more compliant Pedro, carefully established his second son in a quiet boardinghouse. With the multiple goals of occupying the older boy's idle time, preparing him for a trade in case his intractability persisted, and removing him from possible influence over Pedro, his father apprenticed Cajal to a barber. From his own experience, Don Justo knew that there was more than one way to produce a doctor. The joy Cajal had found in Victor Hugo and Velásquez sank beneath the lathering of beards and the gossip of the barbershop. Neither was Cajal encouraged by his master's firm assertion that he should cheer up, because "within a short time you will rise to be a fully qualified barber and will enjoy the wonderful pay of three dollars a month in addition to the tips." When this experiment also failed to cure the apprentice of his passion for drawing and his neglect of academic work, Cajal's father ultimately removed him from school altogether and apprenticed him to a shoemaker!

After a year of sewing soles on new shoes and repairing boots, Cajal begged Don Justo to return him to the Institute of Huesca. This time, he was more prepared to appreciate his opportunities. A classmate remembered later that while he had "not had any master," he was even then "a manufacture of his own individual powers, of his strong will, and of his outstanding intellect." Soon he was not only reading Greek and Latin but also successfully studying the physics of Hermann von Helmholtz, mathematics, astronomy, and natural science. He remembered seeing the eclipse of 1860 and vanquished an aversion to geometry and trigonometry when he realized that these tools had allowed the Greeks and then Copernicus to describe the heavens. To his delight, he discovered the insect world of J. Henri Fabre, the great French naturalist. Finally, in 1869 at the age of seventeen, he finished his secondary education at about the conventional time.

Don Justo, wishing to make certain there would be no recidivism, packed the new graduate off to Zaragoza for a brief course of preparation before he started medical school.

Most of the class work was didactic, but Cajal discovered an attraction to anatomy, in part because he was so good at drawing what he dissected. When the dean of the Faculty of Medicine appointed his former classmate Don Justo to the Anatomy Department, the whole family moved to Zaragoza. Father and son dissected side by side. This work continued throughout Cajal's entire medical school career, and finally his father exulted in his son's ability to create anatomical watercolors and drawings both beautiful and accurate. "My pencil," he wrote, "which was formerly the cause of so much bitterness, at last found grace in the eyes of my father."

Although already devoted to anatomy, by his own admission Cajal was hardly the star of his class in other subjects. Aside from dissection, his interests outside the medical curriculum tempered a complete devotion to the process of becoming a doctor. He remained a weekend naturalist, read novels and philosophy, published poems in the local papers, and—infected by "an epidemic of lyricism"—became infatuated with gymnastics. In his persistently quixotic way, he made an unsuccessful attempt to fall in love during these years. The effort failed because, at twenty-one years old, he found women incomprehensible—a fact that, to his credit, he recognized. With some pride in his own accomplishment, and to a delight tempered by surprise on the part of his father, in June of 1873 Cajal graduated from the University of Zaragoza with the title of Licentiate in Medicine.

THREE

A Laboratory
in the Kitchen

EARLY IN THE ACADEMIC YEAR of 1884, Professor
Santiago Ramón y Cajal stood before his class of medical students at the University of Valencia, charged with the task of
teaching them normal and abnormal histology. This was a
subject of little interest to those particular students, and
indeed to most of the other members of the faculty and the
small scientific community in Spain. As a medical student a
dozen years earlier, Cajal himself had been more indifferent
to histology than to almost any other subject in the curriculum. His own excursions into the world of the infinitely small
had begun during the doctoral examinations he completed
in Madrid after finishing his studies at Zaragoza.

These two events had been separated by a brief career as a
medical officer during the first Cuban rebellion against
Spain. Service in the army had not been Cajal's idea. The
insurrection in Cuba resulted in "the so-called *draft of Castelar*" named for the republican president who ordered military suppression of the ungrateful Cubans, wearied of a
colonialism that encouraged slavery, corruption, and brutal-
ity. Only twenty-two years old when he arrived in Havana, the

inexperienced doctor initially labored to take care of three
hundred very ill patients. Cajal had only symptomatic treatments to recommend for the soldiers, felled one after
another by smallpox, dysentery, tuberculosis, and malaria.
The mainstay of his pharmacy was quinine, and little else.
Like all the other practitioners during this decade just prior
to the discoveries of Pasteur and Koch, Cajal and his colleagues in Cuba had little understanding of infectious disease. Often the miasmas of the swamps were invoked to
explain epidemics. Cajal's quarters were adjacent to the ward
that harbored all this contagion, separated by only a few feet
and a flimsy partition. Since it had not yet been discovered
that mosquitoes spread the disease, no one took precautions,
and as a result, the doctor soon fell seriously ill with malaria
himself. His spleen enlarged, his color turned yellowish, and
he became alarmingly anemic. Then he developed dysentery.

Weeks of recuperation, augmented by "heroic doses of
quinine," allowed him time to study English and to discover
the enthusiastic corruption and fraud in the military hospital
kitchen. The cook and nurses efficiently conspired to steal
and resell food, depriving the sick soldiers of badly needed
calories as they enriched themselves, apparently with a nod
from the commanding officers. Cajal himself doesn't directly
say that the commanders were in on the graft, but one of his
biographers claims that after he "confronted the offending
officers and the cook with this accusation" he was adamantly
advised to ignore it.

Thieves protected by officials have little incentive to
abandon their occupation. Cajal's mild effort to stop the

larceny earned him nothing more than an attack by the major in charge of his unit. On the night this same officer ordered the stabling of his horses inside the ward to keep them from being stolen, Cajal threw the beasts out. For this he was charged with insubordination, an indictment given weight by his proclamation to the major that he, Captain Cajal, was in charge of the hospital and that it would not be turned into a filthy stable. Shouting "Who are you to disobey me? I am a representative of the supreme authority here," the furious major threatened Cajal with prison. The supreme authority meant the commander's uncle, an important general in the army staff office. The general's influence was not enough to prosecute the case, however, strengthening the charge of collusion, and it was dropped. But in these events Cajal had been exposed to the Carlist officers as a liberal, and they went out of their way to annoy and bully him.

Though he liked Cuba and admired the native Cubans, he soon grew to hate military service and was disgusted by this firsthand experience of Spanish colonialism. So vigorous grew his republicanism that he called the Spanish occupation of Cuba a "catastrophe" and was horrified by "the tremendous blunders of our overseas policy." Eventually, cynical and malarial, he became unfit for military service. Not only was he unfit; he was also destitute. Corruption in the army had spread to the paymaster general, "who absconded to the United States with ninety thousand pesos and a strumpet." Without a paycheck for many months, Cajal finally wrote asking his father to send money for the passage

home, and in May of 1875 he sailed for Spain to prepare for
his doctoral examinations.

He managed to pass those exams with a doctor's degree
and a headache, but the real magic of his sojourn in Madrid
had been an exposure to a friend's good microscope, and "the
amazing spectacle of the circulation of the blood." He had
watched the flow of corpuscles through tiny blood vessels in
a living frog's webbed foot. This moment seems to have
ignited all Cajal's imagination, combining his intellect with
his highly developed visual sense, and propelled him toward
a new view of the natural world. Jules Verne, whom he had been
excitedly reading, gave way to Charles Darwin and Thomas
Huxley. Very quickly he began to leave behind the parochial
Spain of the Inquisition, a country that still doubted its own
scientific legitimacy. Cajal occasionally scolded his contemporaries for concealing their laziness behind a belief that scientific discoveries "are gifts from God—gifts generously
bestowed by Providence on a few privileged souls invariably
belonging to the hardest working nations, in other words
France, England, Germany and Italy." For Cajal, his country
was intellectually behind its time. Ahead of him, down the
eyepiece of a microscope on a glass slide, lay the unknown
civilization of the tiny.

When he went back to Zaragoza as a faculty assistant,
Cajal immediately scraped together enough money to buy a
serviceable French Verick microscope of his own (costing
him an enormous sum in 1877, equivalent then to one hundred and forty U.S. dollars, a debt that required four pay-
ments to conclude). With it he acquired a serviceable Ranvier

microtome for cutting thin sections of tissue. He knew almost nothing of the techniques involved in either making slides or looking at them, and was just as ignorant of the primitive staining methods then available. He had no teachers but books. However, with these basic tools and an inquisitive, obsessive mind, he began—in the attic laboratory he constructed under the eaves of his Zaragoza rooming house—to look at blood, skin, muscle, and nerves at magnifications that reached eight hundred times.

Years later in his book *Advice for a Young Investigator*, Cajal wrote with Iberian passion about observing and recording what he had seen: "As with the lover who discovers new perfections every day in the woman he adores, he who studies an object with an endless sense of pleasure finally discerns interesting details and unusual properties. . . . It is not without reason that all great observers are skillful at drawing." For Cajal the microscope embodied both science and art, allowing a compulsive and passionate contemplation of the unknown, while his synthesizing mind interpreted what he was recording in meticulous drawings. Later, when roll film and better cameras became readily available, he was able to take pictures through the microscope, producing photomicrographs.

Though he read German poorly and the scant English picked up in Cuba even less well, Cajal intensely studied the histological methods described in the leading German, English, Italian, and French journals. Nothing sufficiently sophisticated was being published in Spain to teach him what he needed to learn. In fact, even in the latter half of the

nineteenth century almost all Spanish anatomists still considered histology a kind of fantasy. Concentrated as they
were on surgery, they devoted themselves—and their students—to gross dissection.

Neither were most Spanish scholars and researchers at all
disturbed by this indifference to the minute. When the great
German histologist Albert von Kölliker stopped to inspect the
Museum of Natural Science in Madrid, he was ceremoni-
ously shown their splendid French microscope, still housed
in its immaculate mahogany case—never opened. The director confided to Kölliker that he didn't know how to use the
instrument, and therefore, out of respect for him, neither did
anyone else at the museum. Before too many more years
passed, Cajal, using much less grand equipment, had made
himself one of the best microscopists in the world.

But this ability was not without significant cost both to
himself and what would soon become a growing family. Playing chess with his lawyer friend Francisco Ledesma in a cafe
late one night, Cajal suddenly began to cough up blood.
Quickly things got worse, and a frothy pulmonary hemor-
rhage threatened to obstruct his airway. He called his father,
who immediately made the correct diagnosis of tuberculosis,
probably contracted at the same time as the malaria that had
felled him in Cuba. Over the next few months, Cajal—who
knew enough to be worried—became so morbidly depressed
that his father sent him, accompanied by one of his sisters, to
take a cure at the baths of Panticosa. Whether it was these
waters or Cajal's "system of treatment [which] consisted of
doing everything contrary to the advice of my doctors," he

recovered after a few months. He returned to Zaragoza as the director of the Anatomical Museum, subdued and hardened by illness into a new maturity, determined to marry.

More sure of himself now, and definitely much more sophisticated, Cajal was attracted to a young woman he happened to meet one day strolling through the city with her mother. Although he didn't immediately recognize the slender, modest, and beautiful Silvería Fañanas García as the child he had once terrorized on the school steps in Ayerbe, he did notice the same "great green eyes framed with long lashes, and the luxuriance of her fair hair." Disregarding his family's protestations that they were both too young and too poor, the pair was secretly married in 1879. They established their first household on his meager salary of twenty-five dollars a month (plus another six or eight dollars each month earned by tutoring) amid whispers that the new husband would either fail entirely or soon be dead of TB. Throughout their many years together, Cajal credited his wife for much of his success. As the "spiritual director" of their family, she took charge of the household and supported him in his scientific life without complaint that such work provided only enough income for a relatively simple life. Married and employed, Cajal now seriously began to apply himself to his microscope.

In addition to his natural gifts, he intuitively understood the scientific method from his first attempts to study microscopic anatomy. He not only manufactured all the specimens himself; he also tediously investigated and then, in drawings both beautiful and precise, recorded similarities, differences, and patterns. By the time he successfully competed for a position on the Faculty of Medicine at Valencia late in 1883, he had become not only the father of three children but an accomplished observer of what most others could not see or didn't understand, as well.

Cajal's laboratory in Valencia occupied a table in the kitchen of his own house, and after meals his wife and children hurried out. Dishes were replaced by the scientific tools he had acquired five years earlier: the microscope, a simple microtome that incorporated a barber's straight razor retained from his apprenticeship but now employed as a cutting blade, and a few vegetable and aniline dyes in glass containers. With only this basic equipment, by 1884 he had made himself a scientist with some slight recognition in his native country and had even published papers in Spanish detailing his findings. Only a few of his countrymen read them, though, and no scientists outside Spain knew they, or their author, existed. By his own later admission, these early efforts were "pretty weak," but by then he knew the techniques even if he was uncertain exactly what he was trying to find.

Then suddenly in 1885, an epidemic of cholera struck Valencia, and as a result Cajal momentarily abandoned his histological studies, but not his microscope. Pasteur, Lister, and Koch were defining microbiology at almost the same time, but the new science of unseen organisms was not universally accepted by medical school faculty members, practitioners, or the public. Great debate erupted as to the cause of the local epidemic, and the population, urged by excited comment in the local papers, argued about whether microbes were simply a fiction invented by those who thought up treatments in order to make money. The older doctors clung to the miasma theory, while newer graduates (like Cajal) recommended boiling the drinking water to kill the organisms. In the midst of the general confusion, alarm, and increasing death rate, a noted physician arrived with a supply of live anticholera vaccine, which only inflamed the debate. With the epidemic at its height in July, several people died on Calle de Colón, the street where the Cajal family lived. It must have terrified them further when, at this moment, their fourth child was born.

To help calm the growing chaos, the provincial government of Zaragoza asked Professor Cajal to study the disease and to determine whether it was caused by a noxious vapor or a germ. They hoped he could determine if the outbreak really was cholera and recommend the value of various preventions and treatments. Turning his microscope to bacteriology, he was able to confirm that the epidemic (by then spreading throughout Spain) was actually caused by the cholera bacillus but was unable to prove any effectiveness from immunization by the injection of live organisms. By that time, more than fifty thousand people had already been injected. In September, he published a simple method for staining the bacillus and was probably the first to show that the illness can be prevented by immunizing patients with a heat-killed vaccine. Most of the attempts to immunize animals against cholera had been experiments done on guinea pigs, creatures that will produce antibodies against live injections of the bacillus but don't become ill after ingesting it.

Cajal therefore realized that this guinea-pig model could not be used to prove Koch's postulates and establish the infectious nature of cholera. He did, however, prove that injection of the killed vaccine was effective in humans. Had this bacteriological work been recognized beyond Zaragoza, Cajal might have been tempted to abandon histology in favor of the then more glamorous (and certainly more remunerative) study of the microbe. But it wasn't, and he soon returned to his slides.

Hour after solitary hour, alone at the kitchen table among the remnants of meals, he cut off small bits of organs and prepared the tissue for examination. In modern laboratories most of this repetitive work (later to be called "scut" by generations of medical students) is done by technicians or students, and more and more of the tasks are now automated. Without the speed and precision of mechanical nucleic acid sequence analysis, the human genome project, for example, would still be an ambition.

Cajal had no such luxuries or assistants. He collected whatever tissue he could gather from a morgue or a butcher shop—a piece of cadaver kidney, an ox eyeball—wrapped it up, and took it back to his own house. He dropped the tissue into a dehydrating solution, alcohol, or another fixative, to remove the water from it. Over a wood stove in the kitchen, he heated paraffin until it melted. Into the liquefied paraffin he dropped the bit of muscle, kidney, brain, or nerve and left it to harden, the wax both filling up empty spaces in the tissue and holding it firmly together. With a knife, he next trimmed away the excess paraffin until only a small face of tissue was just barely exposed on one side, before he mounted the block on the stage of his microtome. Each turn of the instrument's large round crank automatically advanced the stage with the tissue a few microns closer to the razor he used for a blade. A ribbon of paraffin accumulated on the knife's edge, each piece containing a translucent section of tissue only a few thousandths of a millimeter thick. Choosing the most perfectly cut of these sections from exactly that part of the specimen he wanted to examine, Cajal picked them up with a fine camel-hair brush and placed them on a clean glass slide washed in acid. He heated the slide over an alcohol lamp, melting the paraffin just enough to thoroughly expose the tissue and at the same time sticking the tissue to the glass.

For coloring, he initially had only the few simple stains then commonly in use. His cluttered laboratory shelves were stocked with glass-stoppered bottles filled with the blue of hematoxylin and its counterstain, red-orange eosin, and even a few vegetable dyes such as saffron. Carmine had already been used to study nervous tissue and was followed by the discovery of other aniline dyes. By this time histologists had realized that examining fresh specimens introduced so much artifact (false appearance to the structure of the tissue because of the way it was prepared) and provided such scant contrast that little was gained from the method. Tissue that had been fixed and stained revealed cells with much greater clarity and less artifact.

Making the slides required several steps to fix and color the tissue, and in order to reproduce his results, even these methods demanded careful control of conditions. A solution just a few degrees away from its ideal temperature, a few extra minutes in the alcohol, the wrong concentration of a dye, an imperfectly sharpened microtome blade, and the specimen wasn't worth saving. But even the most painstakingly prepared slides often overwhelmed the observer because so much dye was taken up, especially by brain and spinal-cord tissue, that the slides flowered with too many cell bodies and fibers to sort out.

After the slides were made, the real work began.

On many nights, however, it didn't begin immediately. Exhausted by preparation of the slides, he sometimes needed a break before he could return to the specimens he had created. Shortly before midnight, Cajal often put aside the chemicals and microtome, picked up his chessboard, and left the house to his sleeping family. Often he met a friend at a nearby cafe, still crowded late into the Spanish night. Over the next few hours he played chess heatedly before going back home to bed, but not always to sleep. As compulsive as he was about his slides, he was equally obsessed with the ancient game of chess. Once in bed, he would often struggle to find solutions for those games he occasionally lost. On a particular night after returning home to bed, he tossed, turned over, sat up, and exclaimed to himself, "I am a fool! I had a checkmate at the fourth move and did not see it."

Unable to sleep, and with no partner at that hour upon whom to try out his newly discovered chess solution, he returned to his microscope and spent the rest of the night studying and roughly sketching the minute details of a

Cajal (left) playing chess, a game that obsessed him, in the town of Miraflores de la Sierra. This photograph was taken by one of his children in the summer of 1898. *From Cajal, Recollections of My Life.*

cerebellar neuron. Just before the sun rose, he went back to bed with a drawing board and a single sheet of the best paper he could afford. Using the finest nib of his pen, he began to draw what that cell had revealed to him overnight. But the India ink he used dried quickly in the pen, and he continually had to clear the nib by flicking it with his finger, a habit that produced ink spatters all over the bedroom walls. When his wife got up in the morning, Cajal had to explain the splattered, darkened wall to her— a conversation they would have again and again for their entire married life.

FOUR

Cells, Fibers, and Networks

THE END OF THE Napoleonic Wars in 1814 brought to Europe a relative political and economic tranquillity just at a time when the intellectual and technical climate favored scientific discovery. Charles Lyell began to push back both the flood and the church when he published *Principles of Geology*, dating the age of the Earth in geologic rather than biblical time. Lyell both anticipated and encouraged evolutionary theory, and his book traveled in Darwin's trunk on the *Beagle*. Arguing more from intuition than from data, Immanuel Kant had divined that galaxies were masses of stars. Soon thereafter, William Herschel and his son John built telescopes monstrous enough to discover Uranus (though it was at first thought to be a comet) and to make credible observations about the Milky Way and planetary nebulae.

Not only was man finding his place in the universe, but at the other end of the scale histologists were beginning to see the individual units that make up all of biology. Better microscopes, improved methods for tissue fixation, and staining allowed a closer observation of the microscopic. Although the theory that animals and plants are made of cells had been suggested by the mid–seventeenth century, it languished in the absence of broad application to biological understanding. Cells were so poorly seen during the period when anatomists were making their way from gross dissection to the examination of specific tissues that their function was obscure to the professors who still hoped they could extrapolate physiology from cadavers. What the idea of cells suggested for the nervous system was speculation in 1781, when the Italian naturalist Abbott Felice Fontana wondered about the structure of nerves. "I wish to know what the primitive structure of the nerve is," he wrote, at the same time complaining that anatomists had debated the problem for more than three thousand years, "and seem during this time to have done nothing more than multiply doubts and hypotheses." But during the first few decades of the nineteenth century, when Robert Brown discovered the cell nucleus, Rudolf Virchow defined cellular pathology, and Pasteur, Lister, and Koch explained infection, cells were taking their place in the broader context of human and plant biology. By the time Napoleon was permanently carted off to St. Helena, cell theory was a working hypothesis. After 1837, it was a generally accepted part of the biological zeitgeist, at least in Germany and much of Europe.

Yet even before individual cells could be identified, a striped pattern of layers in the cerebral cortex had been recognized, and anatomists knew there was a structure to the nervous system that varied in its different parts. These observations themselves, first made on fresh, unstained tissue crudely cut by hand, were no less murky than their meaning.

As a medical student in 1782, the Italian Francesco Gennari first found a pale line in the cortex of the occipital lobe that bears his name. Better details of this laminar arrangement in the cortex had to wait almost a hundred years, however, for the Russian Vladimir Betz to describe the five cellular layers. But in the intervening time, there was no grand theory to unify the significance of these observations, so each school of anatomical thought joyfully attacked every other school in defense of those structures—real or imagined, actual or artifact—they were all certain they had glimpsed. Often, because of the great variation in methods used for preparing and looking at the tissue, scientists could not find the same configuration twice, much less find validation for their own claims in journals that published similar results from other labs.

In the midst of these debates, the very tiny suddenly became much more visible. The first compound microscopes were made with double-convex lenses that distorted both the shape and color of the objects observed. Improvements began to appear shortly after the English polymath William Wollaston described the "Wollaston doublet" in 1813. This technological advance resulted from the combination of two lenses, flat on one side and convex on the other, separated by a prescribed distance and a diaphragm that limited the field of view. The new optical arrangement produced images freer of spherical distortion but did not eliminate the rainbow of color around the edges. By 1830, Joseph Lister solved this last problem by combining lenses made of different kinds of glass, eliminating the refractive errors and therefore the color aberration. Lister looked at slides of brain tissue with his new instrument and was able to guess at the presence of individual cells. Scientists could now see their slides unobstructed by a fog between the eyepiece and the objective lens. The view down these new microscopes began to reveal details previously unseen, though not entirely unsuspected.

Philosophers had certainly imagined how the brain *ought* to be arranged, and several anatomists thought they had found both globules (cells) and fibers (axons). The better microscopes and techniques permitted enhanced resolution, more contrast, and a better spatial separation of neural elements. By 1836, Jan Purkinje and his student Gabriel Valentine were able to claim that "the entire nervous system is made up of two elementary basic substances, namely the isolated globules [cells] of the covering substance [brain cortex], and the isolated, continuous primitive fibers [axons]." What the relationship was between the cells and the axons wasn't clear to them. Only two years later, Robert Remak attached cell-body firmly to fiber, thereby ending centuries of unresolvable debate. Remak, a Polish anatomist, summoned every bit of available technique and perseverance when he made these observations using his rudimentary equipment, and even then he admitted it was very difficult. He never seems to have doubted that "the fibers originated from the very substance of the nucleated globules [nerve cells]." But this solution to one puzzle only prepared the way for a greater controversy about how these cells themselves might connect, an intellectual war to be fought for the next fifty years.

At the University of Pavia in Italy, Professor Bartolomeo Panizza was investigating the anatomy and function of cranial nerves, and seeking to identify certain areas of the brain that housed unique functions like speech and vision—cerebral localization—at the same moment Remak found that axons were extensions of neurons. Into this environment entered the squat, unprepossessing medical student Camillo Golgi.

Golgi, like Cajal the child of a humble and obscure physician, was equally brilliant and almost as compulsive, though he had a far less engaging personality. Because the intellectual advances came through Germany and France to Italy faster than they reached Spain, however, he had a slightly easier time becoming a scientist. Nine years older than Cajal, Golgi received his medical degree in 1865, shortly after the Spanish youth blew away the garden gate with his cannon and went to jail. Both became medical students expecting to follow in the steps of their doctor-fathers. Through a combination of genius and luck, Golgi and Cajal developed into driven and intensely curious men each of whom, initially lacking real laboratories, began his scientific investigations in his own kitchen. Golgi did have an advantage during the middle of the nineteenth century, because even kitchen science in Italy stood on much firmer ground than it did to the west in Spain.

For generations, the Golgi family had lived about forty kilometers south of Milan, along the Ticino River in the Pavia region of northern Italy. In 1838, Alessandro Golgi, Camillo's father, graduated from the University of Pavia Medical

School and immediately married his cousin Carolina. Although doctors were few, so were the opportunities for making a living from the medical trade in Italy just then, and the newly qualified Doctor Golgi was forced to accept a job as a municipal physician in a remote province.

The village of Corteno in Lombardy lies in that part of the Alps separating northern Italy from Switzerland. Carolina and Alessandro Golgi had four children there, three boys and a girl, who was the youngest. Camillo was their third son, arriving in the summer of 1843 in the same house where all his siblings were born.

Unlike Cajal's spotted page, Camillo Golgi's school record was exemplary. He excelled in mathematics but also led the way in Latin, Greek, Italian, and natural history. Even his penmanship was recognized as "handsome," and he was never absent. Not surprisingly, he finished his primary courses with high honors, first in his class. This performance not only brought him distinction but also enabled him to move back to Pavia (with his mother and siblings) so that he could finish his secondary education at the Imperial Royal Grammar School, though his father was compelled to remain in Corteno for another two years, taking care of the sick.

The absence of his father at this critical time of his life may have contributed to his later need for recognition; however, even as a child Golgi seems to have been withdrawn and uncertain. Although his father, Alessandro, was officially too old, he eventually asked special consideration to apply for the job of associate physician at the Hospital for Incurables of Abbiategrasso near Pavia. When he was at last selected for the position in May of 1858, he could finally leave the life of a village doctor and be reunited with his wife and children.

The regathering of the Golgi family into a single household coincided with the Resurgence, a dramatic period in Italian history that ended with the expulsion of Austria from Lombardy and the unification of Italy. This period of intense nationalism, not just in Lombardy but throughout Europe, ignited scientific investigation in Italy and pushed aside the last vestiges of Aristotelian doctrine. Scientists and philosophers cast aside metaphysics and speculation about the natural world in favor of what could be seen or measured. Their positivism was ultimately derived from Auguste Comte—but tempered by German science, especially organic chemistry—and supplanted philosophical musings over the relationship between body and soul. Darwin had just published *The Origin of Species*, and although *The Descent of Man* did not appear for another dozen years, Thomas Huxley was already busy promoting radical evolutionary ideas. Professors at the University of Pavia, including the anatomist Panizza, the brilliant but quixotic neurologist-psychiatrist Cesare Lombroso, and later the histologist Giulio Bizzozero, firmly abandoned theology and metaphysics in favor of observation. From the resurrection of Rome and in the shadow of the Papal States, Italian science was vigorously reborn, just as seventeen-year-old Camillo Golgi entered medical school.

As a student, Golgi was dazzled by the intellectual and charismatic Lombroso who, while actually less an experimental scientist than an anthropologist interested in mental illness, sparked Golgi's curiosity about the nervous system. Later in his career, Golgi cultivated many interests, but when he finished medical school in 1865 he seems to have aspired only to the position of "secondary physician," the lowest rung on the Italian medical hierarchy. During that same year, Lister's discovery that carbolic acid is an antiseptic, Mendel's description of the laws of heredity, and the great cholera outbreaks around the world all seem to have escaped Golgi's notice. In spite of his academic brilliance, he still perhaps thought of himself as a village boy.

Golgi wandered: into an internship, into the military, into public-health service during a later cholera outbreak near Pavia, and finally—in search of better pay and less hazard— back into a job as assistant physician at the Hospital of San Matteo. There he again met Lombroso (who, although committed to observation, retained an interest in the paranormal). He began a collaboration with his former teacher, doing autopsies and histopathology examinations of the brains of mental patients. Soon he discovered that, while Lombroso professed to be a researcher, his methods were crude and more often than not his generalizations went untested. Golgi recognized the limits of his mentor and went in search of a more sophisticated, if less colorful, teacher.

The medical school at Pavia had a long tradition in anatomy beginning with Antonio Scarpa (whose name is attached to several anatomical sites) and continued by Golgi's teacher Panizza. The rising star of experimental histology, though, was Giulio Bizzozero who, at the precocious age of twenty-one, became Professor of General Pathology

and directed the Laboratory of Experimental Pathology at
Pavia. Although three years Golgi's junior, Bizzozero had a
firmly established laboratory already producing original
research by the time Golgi joined him in 1867.

In fact, young scientists were powerfully attracted to the
modern, stimulating work being done in the laboratory, and
they flourished under the exciting direction of Bizzozero.
These pathologists were not simply examining cadavers. One
of the new medical graduates was Nicolo Manfredi, who
came from an aristocratic family with money enough to support their son's research into subjects that included the possible therapeutic effects of music. The cultured Manfredi, the
provincial Golgi, and their other contemporary, a pathologist
named Benedetto Morpurgo, bonded to the microscope and
to one another. More than fifty years later, after he had succeeded to a chair once held by Bizzozero at the University of
Turin, Morpurgo remembered their teacher and Golgi working together in that laboratory:

> It appeared as if nature had created them for the sake of
> contrast: the former being tall, slender, agile, confident,
> and almost imperious; the later being stocky, slow, appar-
> ently uncertain and subdued . . . Golgi did not give the
> impression of having a bright future before him. Only Biz-
> zozero perceived immediately that in this quiet, modest,
> and somewhat awkward son of a municipal physician
> from Corteno in Valcamonica, there was a deep faith and
> bold power; the seriousness of intents made it possible for
> such different souls to understand each other, and favored

a collaboration that would lead to the resurgence of Italian biology.

The investigation of tissues and their pathological alterations had shown the close connection between fine structure and function, and had demonstrated that the natural sciences could no longer be divided between those that investigated morphology and those that speculated on the phenomena of life.

This laboratory with an imposing name was in fact nothing more than a small spare room equipped with four microscopes, a few dissecting instruments, chemicals for making stains, and other reagents. For the purchase of supplies and equipment, the director received the pitiful sum of four hundred lira (about eighty dollars) each year. But even with these limitations, the laboratory produced original work in several areas, including new studies on the blood-producing function of bone marrow and the structure of the retina. Golgi's own early investigations were merely·clinical accounts of patients with various mental disorders, but as he distanced himself from the influence of Lombroso, he too discovered how to produce original research.

Golgi began by reading Rudolf Virchow's *Cellularpathology*, published ten years earlier. In this text, the great progressive German medical reformer democratically described the body as "a cell-state in which every cell is a citizen," and that disease is "merely a conflict of the citizens in this state, brought about by the action of external forces." Even though he was writing as a pathologist, Virchow had a great deal to say about what he

termed the *"structure and arrangement of the nervous system."*
He began his discussion in this section of the textbook by
admitting that both the anatomy and physiology were entirely
uncertain, and went on to say that the view of contemporary
neuropathologists "is a very erroneous one."

> For they fancied they saw in the nervous system an unusu-
> ally simple whole, from the unity of which resulted the
> unity of the body in general, of the whole organism. But
> even though one has nothing but very rough anatomical
> ideas concerning the nerves, still one ought not to shut
> one's eyes to the fact that this unity is in a very sorry plight,
> and that even the scalpel demonstrates the nervous sys-
> tem to be an apparatus composed of an extremely large
> number of parts of relatively equal value. . . . and in great
> measure independent of one another.

This assessment was not at all encouraging to a young scientist. Nonetheless Golgi began to experiment with various
fixatives and stains on slides of brain tissue in an effort to see
the nervous system without distortion, to catch it off guard,
as it were, and flush it from hiding. Although Purkinje had
used a primitive microtome and employed crude methods to
both fix and stain tissue, most preparations were still simply
unhardened tissue sliced with a razor, squashed between two
glass slides, and examined fresh.

Golgi knew of fixation techniques, as well as staining with
carmine, other aniline dyes, and gold chloride, but all these
methods had limitations when applied to the nervous system. Alcohol could be used to harden most tissue, but it dis-

solved the protective myelin coating around axons and so was by itself unsuitable. Another good fixative, osmic acid, was far too expensive. Golgi began to harden his specimens and to prepare them for cutting and staining by submersion in potassium dichromate, a reagent he chose because it was cheap, readily available, and preserved fresh tissue well. Later this fortuitous step helped prepare the way to his most revolutionary scientific discovery.

The available staining techniques themselves often concealed unexpected traps. Artifacts, those findings not present during life but produced by the method of study, were easily introduced, confounding observers. These phantoms might result not only from poorly understood effects of the reagents themselves but also from something as unnoticed as minor variations in conditions—the time tissues were submerged in a reagent, the concentration of a solution or its temperature, for example—when preparing the slides. Just as often in examining the nervous system, though, too many cells were stained and histologists were overwhelmed by the enormous number of things to look at on the microscope stage. While they might be able to see thousands upon thousands of cell bodies, it was impossible in this forest of superimposed cells and crossing fibers to find where or how these objects connected. Things simply disappeared into what the greatest German anatomist of the preceding generation, Jacob Henle, had termed an "amorphous, finely granular interstitial substance." This substance was, he claimed, without cells at all.

Histologists had already learned to differentiate distinct

kinds of cells in the nervous system. They also had a primitive understanding of a difference between the fibers (axons and dendrites) that Remak had proven were attached to the cell bodies. But the precise structure and organization of these globular cell bodies and their accoutrements were out of reach. In 1870, no one knew what a neuron really was, much less what it looked like or how it worked, and therefore any attempt to describe an accurate three-dimensional structure of the nervous system was impossible.

Golgi wisely began his investigations of nervous tissue by looking at something relatively simple. His first study of the brain reported not on the architecture of cells themselves but on the spaces around the blood vessels and the pia, a membrane plastered over the surface of the cortex. Only after he had proven that some of these spaces were really artifacts resulting from shrinkage of tissue during fixation did he venture deeper into the substance of the brain. Thanks to Virchow and his student Karl Deiters (who, before he died of typhus at twenty-nine, produced an enormous amount of pioneering work), it was already accepted in Germany that small, round cerebral cells without an axon were not nerve cells but rather neuroglia—cells that have turned out to be the glue, insulation, and janitors of the nervous system. This theory of a cellular cement was, naturally enough, opposed by the proponents of Henle's notion of an "amorphous matrix" upon which neurons were supposed to develop, but which was otherwise thought to be a sort of cellular goo, without any structure.

In a manuscript found after his death and published in

1865, Deiters had accurately drawn and started to classify the star-shaped neuroglial cells. When Deiters's papers were finally published, scientists immediately recognized the work as ground-breaking. Using these drawings as his beginning point, Golgi examined the intercellular substance of brains stained with carmine. He soon discovered that, if he stained only very fresh tissue, the thing authors variously referred to as a "finely granular" or "amorphous" substance vanished, and in its place the actual scaffolding of the connective neuroglia appeared as if exposed by the magic of invisible ink.

Golgi eventually proved that the "amorphous substance" observed by Henle and others was the result of the breakdown of cells after their death, what he called "degradation of the fibrillary substance," that occurred in old, poorly fixed tissue. In fresh specimens, or in well-preserved tissue, Golgi found a more accurate picture of the supporting structure, showing "the elegant shape of the connective cells," and that "the interstitial stroma of the cerebral cortex [the amorphous substance of Henle] consists mostly of the connective cells and their processes." Between 1870 and 1872, Golgi confirmed these findings and expanded on Deiters's classification of neuroglia in papers that began to earn him a reputation throughout Europe as a promising and original young investigator.

Work done by Golgi, Deiters, Remak, Virchow, and others had, by early in the 1870s, provided the tools and understanding to expose the cellular structure of the nervous system. Such a discovery would build not only a three-dimensional model of the brain and spinal cord but might also clarify the emerging theories about their electrical nature. The only thing lacking was a stain that would outline just enough neurons in sufficient clarity to show the basic shape of their cell bodies and to permit fibers to be followed to their destinations, but without exposing so many as to obscure the entire map.

FIVE

The Black Reaction

AT THE AGE OF twenty-eight, Golgi couldn't make a living doing research in Pavia. Poverty and the energetic encouragement of his father, who was always motivated by financial worries, forced him to leave the collegial atmosphere of Bizzozero's laboratory. Ironically, the job he finally applied for was that of chief physician at the Hospital for Incurables in Abbiategrasso, where his father had labored. Alessandro was convinced that if his own assistantship had been good enough for him, then the job of chief physician at the same hospital was even better for his son and ought to bring Camillo some security, if not necessarily happiness. So ignoring Golgi's obvious commitment to scientific research, his pragmatic father "began to bully his son" into seeking an "honorable and stable position." Golgi won the job, and moved to Abbiategrasso in June of 1872.

But he hadn't been in his new post very long before he enthusiastically picked up his prior research and began to make slides. While the Laboratory of Experimental Pathology had been rudimentary, he had his colleagues for support. In his new position he worked alone in similarly bare-bones

conditions at a microscope set up on the kitchen table in his
small hospital apartment. Isolated from collaborators and a
real lab, the new chief physician did not lack imagination and
fervor. There being meager treatment to offer the six hundred
and fifty-five patients considered incurable, his duties at the
hospital that housed them were minimal. He had plenty of
time, no family, and—apart from disputes that soon developed with the hospital director—little to occupy his driven
imagination except histology.

His life as chief physician, while not demanding, was not
without difficulty. Not long after he assumed his duties, Golgi
irritated the director by complaining about the diets of the
patients in the infirmary and chronic wards. He was con-
cerned, Golgi wrote, that while patients were offered enough
daily calories, these arrived mostly in the form of starch,
sugar, and wine, rather than protein. The director, a rigid
bureaucrat, sought to suppress his young chief physician by
enforcing the hospital curfew and locking the hospital gate at
night, but not providing him with a key. Unhappy with the
restriction of his freedom, Golgi revolted and moved out to a
private apartment, a solution that cost the director money.
Lodgings away from the hospital caused further problems at
night because, if a patient became ill, someone had to go and
fetch the doctor. The situation was so unworkable that, eventually, Golgi received a key to the gate and moved back into
the hospital apartment. The kitchen became his lab.

Though he hadn't been offered an academic job at Pavia
(or anywhere else), Golgi had not for a moment considered
abandoning his search for a stain that would more clearly

show him neurons. He tried dozens of reagents without success. The experiments he had started doing with solutions of silver nitrate in Bizzozero's lab—fooling around with various fixatives, concentrations, and durations—while working on the anatomy of the spaces surrounding cortical blood vessels had given him a direction, though. In fact, while searching for the true anatomy of those spaces, he had noticed on the edges of his slides that some of the cortex happened to remain attached to the pial membrane. Golgi was too observant to miss the fact that these slides also showed a few partially stained brain cells. It was fortunate that he again picked up his idea of staining with silver. He must have worked both methodically and intensely, for he soon had the answer.

Golgi never tells exactly how the method came to him, and although he soon wrote about his new technique, he was slow to publish the precise details. Finding the proper concentrations of these already known reagents and, just as important, the proper sequences and durations for soaking pieces of brain to make the stain work was clearly trial and error. But, as Pasteur famously observed, "Chance favors the prepared mind," and at this moment in his career Golgi was eager to look at the nervous system in novel ways. Writing earlier about the contradictory results from different labs studying neuroglia, he said, "It is clear that these conflicting results must be largely attributed to differences in the reagents and the protocols used." So hour after hour in his small makeshift laboratory, Golgi tried various brews of reagents, concentrations, and timing. By late in 1872, he made the chance discovery that when mature nervous tissue was fixed in potassium dichromate for a prolonged time and then soaked in a weak solution of silver nitrate, a chemical reaction produced traces of silver chromate salts loosely deposited throughout the cytoplasm of some of the cells. When he finally stumbled on the right combinations and found the black silver chromate salts accumulated in only about 5 percent of the neurons, he immediately realized what he had done.

Some neurons were exposed by their silver-impregnated, darkened outlines! Golgi quickly named this process *la reazione nera*—the black reaction. The technique was erratic, however, and early in these experiments his results were difficult to predict or reproduce, much less to write about in detail with conviction.

Years later Cajal himself tried to imagine how his great rival might have felt upon uncovering the black reaction in 1872. Stretched out on the slide were only a few stained neurons and glia—*but the chosen ones were outlined distinctly*. Here is Cajal's romantic picture of Golgi, looking at the entirety of a neuron, from cell body with its dendrites to the length of its axon, something no one else had ever clearly seen before:

> What a fantastic sight! On a yellow, completely transparent background, there appear sparsely scattered black fibers, smooth and small or thick and prickly, as well as black, triangular, star- or rod-shaped bodies! Just like fine India ink drawings on transparent Japanese paper. The scientist gazes upon it in astonishment. He is more accustomed to the chaotic images produced by carminic acid and haema-

toxylin, which yield one dubious interpretation after another. Here, on the other hand, everything is absolutely clear, without any possibility of confusion. There is nothing more to interpret.

In these final two sentences, Cajal's own lyrical imagination may have overwhelmed him. Plenty of possibility for confusion, years of interpretation, and sometimes violent argument lay ahead for both the Spaniard and the Italian, as well as many of their contemporaries. At this moment of discovery, though, the immediate future was filled with hour after hour of tedious work for Golgi, peering through his microscope at wonderfully outlined cells.

Golgi was perplexed as to why only a few of the cells stained black but elated to have discovered a method that exposed naked neurons, stripping away the confusion of both excess numbers of cells and the jumble of faintly outlined cell bodies and axons. On February 16, 1873, he wrote, with some exaggeration, to his friend Manfredi, still at the Laboratory of Experimental Pathology in Pavia, "Delighted that I have found a new reaction to demonstrate even to the blind the structure of the internal stroma of the cerebral cortex." He begged Manfredi to supply him with the reagents—crystals of silver nitrate and dichromate—as well as to send tissue from a just-slaughtered horse. Armed with his new method and with fresh resolve, he worked at a frantic pace. Morpurgo, who had continued to work in Bizzozero's laboratory and always maintained a close friendship with his former colleague, later wrote:

With great emotion I remember that period of Golgi's life. Alone, without funds for his research, without the support of colleagues or the advice of teachers, with the little treasure of his microscope, spurred by the confidence in the importance of his discovery and delighted by the results that he was obtaining day after day, he worked in silence for an entire year.

When he broke that silence publicly, the first of his preliminary papers—published in the *Gazzetta Medica Italiana-Lombardia*—reported a true scientific breakthrough. Like many such original discoveries, however, Golgi's black reaction was not immediately understood. His earliest descriptions of silver staining were, in fact, largely ignored. But the suddenly naked neuron could not go unnoticed for long. Golgi began immediately to use the power of silver impregnation to find new details of cellular anatomy, including the fact that as they moved farther away from the cell body, axons did not always remain a single fiber. After he published examples of this terminal axonal branching, what he called "collaterals of the cylindraxis in the spinal cord," and organized all the discoveries he had made using the black reaction, news of the "Golgi method" spread rapidly through the labs of Europe. He reported so much between 1877 and 1880 that the volume of his papers sometimes smothered the most significant discoveries.

This was especially true of axonal branching, a finding later reported by Cajal who, at the time he found the branches, didn't know of Golgi's prior observations. Branching was an important discovery. It led to conflicting interpretation by neuronists and reticulists and fueled the personal antagonism between Cajal and Golgi, the two main representatives of these schools. In Italy, branching was further proof of the reticulum, but in Spain it was simply another way for axons to approach dendrites.

Between the time he found the method for silver staining and his departure from Abbiategrasso, Golgi became reclusive. When he eventually returned to academics as professor of anatomy, first briefly at the University of Siena and then at Pavia, the notice of these academics gratified him, and he was certainly delighted to give up his duties on the mental wards. But his passion remained silver stains. In his last months at Abbiategrasso, he devoted himself to an exhaustive study of the cerebellum, in part because Purkinje and his students had already made a good start on knowing the cells in this part of the brain. By changing the length of time he fixed tissue when submersing it in the dichromate solution, he found it grew easier to see neurons, glia, and cell processes, or a combination of these elements at the same time. He began to categorize the cells in various layers of the cerebellum, work initiated by Purkinje but still contested then, and to make detailed observations about neurons and neuroglia.

But even the terms *neuron, axon,* and *dendrite* were not yet used in 1874 when the work was published. Writers employed different descriptions of what they saw to explain the same things. Deiters certainly had seen two kinds of cell processes (axons and dendrites), but for him, axons never branched. Golgi slightly modified Deiters's nomenclature by

calling axons "nerve extensions," "axis cylinders," or sometimes "nerve fibrils." He referred to dendrites as "protoplasmic extensions." Deiters's description of a possible third
system of fibrils confused the terminology further. These
third fibrils of Deiters, what he called the "cylindraxes," and
which did branch, were probably nothing more than bare
axon terminations on dendrites (boutons). It was this system
of "cylindraxis" that convinced Deiters that he had found the
network. Golgi later used this same term to mean axon when
he classified neurons into Type I and Type II cells depending
on the length of their terminal branches, a finding that,
according to Mazzarello, "represented a great theoretical
contribution by Golgi to the neurosciences because it made
it possible to elaborate the concepts of local and long-
distance circuitry." The discovery did not, however, lead
Golgi away from the reticular theory but rather strengthened
his belief in it and Deiters.

Regardless of what he called it, it is clear that Golgi saw the
neuron as composed of a single axon exiting a cell body
laden with dendrites. For him, though, dendrites did not wait
like tentacles of a jellyfish for stimulation but somehow nurtured the neuronal cell body. Just as clearly, he believed that
the fibrils of the cylindraxis he found on axonal branches
were all part of a reticular network that connected the nervous system. These tiny fibrils were, Golgi thought, the lace-
work that hooked up every cell to every other.

Reasoning that a purely sensory phenomenon such as
smell might have a unique microscopic anatomy, he turned
to the olfactory system. Here, he believed, were the fine, ter-

minal axon-to-axon collaterals that connected all the cells sensing smell. Although Golgi later moved away from a belief in such localization of special functions to specific regions of the brain—and toward a much more holistic view of cerebral organization—at this point he was, like most of his contemporaries, searching for distinctive cellular anatomy to explain specific physiology. Along with these efforts, he sought evidence in support of the reticular theory. His nervous system was a kind of ant colony, made up of passages that routed information through an entirely connected labyrinth.

Scientists found reticular connections everywhere in biology at that moment (Wilhelm His and Purkinje both thought they had also come upon them in the heart), and theories abounded. Deiters, although he could differentiate between axons and dendrites, had clearly proposed neuronal networks, and he was supported by Joseph von Gerlach's early observations using carmine and gold stains. Looking at carmine-stained slides of the spinal cord, Gerlach proposed a double origin for sensory cells entering and motor cells leaving, but he insisted that they were directly connected. In the 1867 edition of his influential histology text, even Kölliker, who eventually converted to the neuron theory, favored a network strung together by protoplasmic processes (dendrites). All these ideas were efforts to explain how sensory impulses enter the spinal cord and produce an effect.

Why, these pioneers marveled, do we so quickly pull away from a flame? Receptors in heated skin surely stimulate a nerve fiber that eventually enters the spinal cord, and somehow there is an automatic withdrawal of the heated hand (a phenomenon now known as a spinal reflex). But exactly how did it work? Even more perplexing to them were the stimuli that entered the spinal cord or brain stem, and magically ascended to consciousness. Contemporary drawings of the spinal cord pictured sensory nerves entering in the proper place and motor neurons leaving correctly, but the central white matter in between was a kind of fantasyland where specific fibers dissolved into a jumble of hope and guessed at destinations and junctions. Gerlach had sought to remedy this failing by proposing a network of cellular connections through "protoplasmic processes" that he believed linked each cell to every other. Though Golgi disagreed with Gerlach's cellular specifics, he firmly embraced the idea of the network. All the Italian's subsequent great work in neurohistology was influenced by this preconceived belief.

Golgi had been very ingenious and productive in his homemade lab at the mental hospital. In addition to the black reaction, he also took up a problem previously investigated by Virchow and others, who thought that trauma to the brain caused calcification inside the injured cells. From prior work with Lombroso, Golgi knew that a variety of calcium salts could be found in diseased brains, and he found a chemical method for distinguishing between them. When he added sulfuric acid to his specimens, some of the deposits produced gas bubbles and crystals that could be seen on the slides, and some did not. In 1876 Golgi published a paper establishing that calcium phosphate could sometimes be found inside injured cells, but that calcium carbonate (which

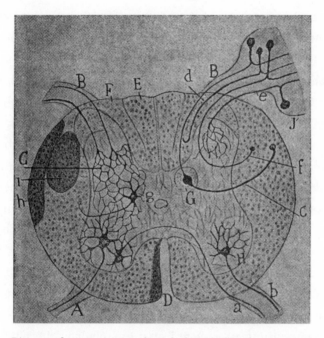

Diagram of a cross section through the spinal cord as imagined before the silver stain. A: motor root leaving the cord to become a peripheral nerve; B: sensory root bringing information to the cord; C: supposed network of connections in gray matter; H: anterior horn (motor) cell body providing the axon that eventually synapses with muscle at the motor endplate; a: incorrectly pictured motor axon crossing in the cord; j: sensory ganglion, showing both its bipolar anatomy consisting of a long dendrite (cut off at right upper) and short axon d. *From Cajal,* Recollections of My Life.

made the gas) appeared only in their membranes, though not always. Trauma did not appear to be a crucial variable. He concluded that the calcifications found in the brain were nonspecific and could be found in many different conditions, including not only injury but also infection, tumor, and congenital abnormalities. This amazing accumulation of laboratory work, and more, Golgi accomplished in the single year of 1873. But it brought him no closer to a professorship.

The Italian academic medical world was as complex as the College of Cardinals. When he had applied for the position of Professor of Histology at the University of Pavia in October of 1873, Golgi wrote to Manfredi, he was found not worthy. On his behalf, well-placed academic friends began to lobby for Golgi at the Ministry of Public Education and among their influential friends in Rome. Even Bizzozero intervened to support his former student. All this intrigue helped, and by the summer of 1873, Golgi's scientific star was rising so fast that Bizzozero warned him to keep his most innovative ideas to himself until he had tenure. In January of 1875, the Minister of Public Education saw to it that Golgi was made a Knight of the Order of the Crown of Italy for his work and then appointed him to a vacant anatomy chair at the relatively minor University of Siena. Whether it was the promise of a competing position, his growing scientific acclaim, or the efforts of his friends, the rector of the University of Pavia suddenly found Golgi the *only* person worthy to hold their chair as Professor of Histology. Pavia was a much more prestigious school, so Golgi only slowed down as he passed through Siena. In 1876, at age thirty-six, he returned to the university where he had been a medical student and remained there for the next forty years.

There was a brief moment just after the move to Pavia when the persistent Bizzozero tried to lure Golgi into joining him at the even-more-prestigious university in Turin. This effort failed, but in the course of negotiations he was introduced to Bizzozero's niece, twenty-year-old Lina Aletti. Golgi was a shy man, not given to emotional display, but he immediately fell in love with Lina. Shortly after their first meeting, Golgi wrote to Manfredi saying he had suddenly discovered "feelings never before experienced." These powerful feelings were apparently reciprocated, because after a few months of engagement, the pair were united in a civil ceremony at the city hall in Varese where Bizzozero lived. Thus Golgi began his long, successful academic and married life. As Lina came to him with a substantial dowry, the childless pair were comfortable, and Golgi was to become not only a scientific hero in Italy but also a successful university administrator and politician.

Ultimately, his most extraordinary contribution would be the discovery that neurons could be stained by silver nitrate, an insight that very soon paved the way for even greater breakthroughs in neuroscience.

THE PERSONAL NOTICE that Golgi sought all his life may have been slow to arrive, but the news of his discovery did percolate through the European anatomical community. Communication among laboratories occurred predictably at both scientific meetings and informal gatherings. Even though distribution was slow, journal reports appeared with regularity, and letters generally arrived at their destinations, so the discovery of the black reaction was soon the talk of interested histologists all over Europe.

Not many technical details about the stain accompanied the news, however. Golgi's 1873 preliminary report in the

Gazette included only a scant methods section, revealing few
specifics about how to prepare and stain the tissue. Golgi
wasn't hiding anything. He really didn't know the best methods yet, a fact he made no attempt to conceal. Even in his
paper published two years later on the fine structure of the
olfactory system, he commented that he didn't know why the
stain worked, or exactly how long to soak tissues in the various reagents. And he admitted that results were still "partly
determined by chance." None of this hedging dampened the
excitement other histologists had for the technique, though,
because when it worked it showed them brain cells in such a
spectacular new way.

By the early eighteen eighties, Golgi had finally organized
all his work—including a much more rigorous and detailed
methodology—and he published *Studies on the Fine
Anatomy of the Organs of the Central Nervous System*. This
effort was noticed throughout Europe and won him the two-
thousand-lira Fossati Prize. His techniques were in wide use
by then, including in laboratories at the best institutions in
France, and anatomists all over the world who studied both
neurological and mental disorders were looking at neurons
by means of the Golgi stain.

In 1887, Cajal paid a call on a Spanish psychiatrist just
returned from France to Madrid. Cajal did not know Doctor
Luis Simarro well, but he did know that he had been studying
at laboratories in Paris and had brought back tissue stained
with Golgi's silver method. And he knew how to make the
slides. The young Professor Cajal had just moved to Valencia,
and was desperate to learn the newest techniques already in

use throughout the rest of Europe. The day he called at the home laboratory of the psychiatrist, he was trying to finish writing a histology manual that he wanted to make as current as possible.

Cajal looked around the small lab and admired several of the newer preparations, but when Doctor Simarro showed him the Golgi stains, he gasped. He was jubilant, in the same way he had been when he first examined the frog's webbed foot. He wouldn't leave the lab, couldn't sleep all night, and far too early the next morning surprised his new friend by coming back to look again at the magic of neurons stained in their entirety with silver. From that moment, Cajal had no other ambition but to understand how the nervous system was organized. He borrowed Golgi's book from the psychiatrist and, in his own kitchen laboratory, sat down to a new task.

Seeing the silver stain focused Cajal's compulsive, intense nature. Even the capriciousness of the Golgi method wasn't a deterrent, as he began to stain the cerebellum of every dead creature he could find. Early in these studies, he couldn't afford experimental subjects any more exotic or evolutionarily complex than the local mice. Immediately he turned this to an advantage. He recognized that the black reaction would enable him not only to study the neurons and glia but also to track down the long axons in thick sections of tissue as they made paths across the slide. Since a mouse brain isn't very big, he could examine the entire organ in a relatively few sections.

For the remainder of his tenure in Valencia, Cajal experimented with the difficult silver stain, varying times of tissue immersion and concentrations of reagents and temperatures, allowing himself but one diversion—the pleasure of photography.

Cajal was an accomplished photographer of both people and landscapes. The artist he might have been had found genuine expression in this craft. Since there was no readily available commercial printing of plates and film was not quite yet commercially available, he prepared all his own materials and developed his own pictures in a home dark-room. From his introduction to photography as a child, he knew the action of developing agents was to reduce emulsions of silver salts to metallic silver in concentrations dependent on the amount of exposure to light. Negatives, therefore, reveal the darkest images where light exposure is greatest because that is the area of the heaviest deposits of silver metal. When a print is made from the negative, these are the parts of the picture that appear brightest. Both his scientific life and his hobby now swam in a sea of silver salts, but Cajal didn't immediately connect them. The joy he took in photography was then simply an artistic diversion and seemed to have nothing to do with his passion for the black reaction.

In the meantime, the chance to take pictures with his friends around the countryside near Valencia attracted him, and he was happy and productive at the university there. But in 1887 a position became available in Barcelona, and he decided that "for a man dedicated to one idea and resolved to devote his whole activity to it, great cities are preferable to small ones."

During the entire first year after he moved, Cajal was

reduced to studying only pathology—an area of investigation he had neglected, now a requirement for the Professor of Normal and Pathological Anatomy. He finished his histology manual (illustrated with more than two hundred original drawings) and amused himself by frequenting the local cafe with friends. Whether it was this stimulation, a rest from the details of his work, or simply the incubation of time, he had a breakthrough. All of what he had studied and been thinking about declared itself when what he called a "new truth. . . . rose up suddenly in my mind like a revelation." By 1888, this truth had been formulated into theories as he began the most productive work of his career.

With the aid of his modifications to the Golgi method and insights about the best tissues in which to search, Cajal set out to prove his suppositions.

SIX   *The Trench of*
                          Science

BEFORE HE COULD APPLY his attention to the complexities of cerebral organization, Cajal now had to rid himself of what he considered a serious vice—playing chess. The
company of his friends in Barcelona was so agreeable and he
had become so good at the game that, while he wasn't losing
any money, he was losing sleep and concentration. Competitive and passionate, he had inherited the same visual memory that blessed his father and now had an uncontrollable
need to use it in the game of chess. He had reached the point
of contesting four simultaneous games, some with his eyes
closed, when he decided that he was in danger of becoming
an old man at a card table competing until he fell off his chair
dead. The game had so addicted him, and so filled him with
a desire for revenge on those who occasionally defeated him,
that he was unable simply to quit.

Ever the scientist, he devised his own cure. He stayed away
from his friends and the cafes. He took no photographs and
didn't even look at his precious slides. But he did make a
thorough study of all the chess books he could find, memo-
rized the most celebrated moves, and at the same time

vowed to abandon a reckless and aggressive style that had characterized his game. For a solid week thereafter, he ruthlessly defeated everyone he could trap into playing against him. Having proven he could do so, both to himself and the competition, "the devil of pride smiled and was satisfied." Cajal left the cafe, went into his lab, and did not sit down at a chessboard again for many years. Exactly how many years is unclear. In his *Recollections* he claims that he "did not move a pawn again for more than twenty-five years." His memory here may have been imperfect. The photograph that shows him playing chess seen in chapter 3 was taken only ten years later.

This liberation from the game returned Cajal to his microscope, but it still didn't mean there was any widely read Spanish journal where he could easily publish what he found there. When he began to produce manuscripts for article after article in 1888, he had to establish his own journal in order to distribute his findings more promptly and, he thought, attract the attention of better-known academic scientists. Still, the famous ignored him. Admittedly he could afford to print only sixty copies of the initial issues of his *Quarterly Review of Normal and Pathological Histology*. At first this journal was devoted only to his own research. Searching for a way to display the value of the work, he immodestly sent them to the principal histologists of the world, whether they asked for them or not. At universities in Germany, Italy, and France, they must have arrived like junk mail.

In his papers, Cajal described an improved staining method he called the "double impregnation" procedure. He began with the Golgi technique by soaking very thin (four-millimeter) pieces of brain in large volumes of a hardening solution containing potassium dichromate, osmic acid, and distilled water. He removed these shavings, washed them, and immediately soaked them in weak silver nitrate for a day and a half. Whereas Golgi would have then cut the tissue, Cajal put it back in a more concentrated solution of fixative for a day or two before soaking it again in silver nitrate for yet another day. The slight technical alteration made to the original method thinned out the visible neuronal cell bodies and fibers (as well as glial cells) still further, making them easier to follow visually. Double impregnation also stained the cells and fibers more deeply, so that the preparations could be made thicker, allowing the longest axons to be pursued on one slide. Tracking down connections in this way produced fewer dead ends and therefore made better maps. Cajal used double impregnation to refine the descriptions Golgi and others had made on the fine structure of the cerebellum.

Technique helped, but it was not the greatest of Cajal's insights as he began his extraordinary adventure into the forest of the microscopic. When he wrote about what he saw during this period, Cajal himself used the simile of roots, trees, and branches to describe neuronal complexities. This was a thicket so dense that, instead of attacking the fully developed neurons directly, he plotted what he called "strategic subterfuges." Many scientists had, of course, recognized the problem of trying to explore the mature overgrown forest, thick with stained cells and fibers, but layered, indistinct, and vague. Some had attempted to simplify the task by teasing out single fibers, as Deiters (and others) had painstakingly done, and some like Golgi hoped to improve their odds by staining with methods that enhanced certain elements while minimizing others. None of this worked very well, although Golgi's "black reaction" worked much better than anything had before it.

Cajal's mind took a different route. He reasoned that if the adult nervous system, completely formed and therefore at its most complex, was too overgrown to figure out, why not look at something simpler? He turned to the embryo.

This was a detour everyone else had missed, but it led Cajal to solutions to many of the puzzles that had confounded generations of philosophers and scientists. It turned out that myelin, the lipid-protein coating around axons—similar in function to the insulation around electrical cords—is the enemy of silver impregnation. Golgi, who had abandoned embryos early in his career and now always worked with adult tissue, produced some handsome specimens, but the most fundamental morphology and organization of the brain eluded him, both because the adult anatomy was so intricate and because myelin inhibited the stain. But in immature embryonic tissue, myelin is scant and the paths less cluttered.

This discovery loosed Cajal's most intuitive creativity. Such frenzied energy and money went into this new research that his wife and children (now five of them) could no longer afford to pay a servant, then considered a necessity in middle-class Spain. While there is scant record of the labor required of Señora Cajal to keep the house and feed and clothe the family, the account of her husband's scientific discoveries during these months is clear. And her loyal efforts were not lost on Cajal, who said of her that "with her self-denial and modesty, her love towards her husband and children and heroic economy, she made my obstinate and dark scientific activity possible."

Later Cajal wrote extensively and passionately about this period of his life, in his autobiography and in *Advice for a Young Investigator.* He remained isolated from the scientific mainstream and had no real teachers of stature and few contemporaries with whom he could debate his ideas, but he doubted neither his own ability nor the authenticity of the scientific revelations that he was now sure his methods would expose. Even though he might be wandering alone in his vast neuronal forest, he knew where he was going and, by then, how to get there. In the rest of the European scientific community, histologists continued to attack problems that were much too big with tools that were still too small.

In constructing his own experimental strategies, Cajal began by dispatching Aristotle. Arguing that the examination of one's own mind for the solution to the problems of natural science ends in the "pursuit of chimeras," he recommended pragmatic use of the senses in observation, description, comparison, and classification. While some of our scientific certainty may have dissolved with postmodernism, before Heisenberg placed his limits on observation and Chaos Theory disrupted prediction, Cajal insisted that the human mind was perfectly capable of knowing immediate causes and invariable relationships. In short, he still clung to a belief in

Newton, and like his father, he preached diligence and hard work. While acknowledging that it helps to have genius, in his book he advised the young investigator to avoid prejudice and maintain self-confidence. Perhaps even there he warned against too much admiration for the authority of great minds, reminding his readers that "great men are at times geniuses, occasionally children and always incomplete." By the time Cajal expressed this opinion, however, he had Golgi's failings as examples, behavior that he didn't necessarily dwell on but could never entirely forget.

Novel experimental strategies, however, convince no one. Cajal next began in earnest to find the evidence. He took issue with Golgi and Gerlach when he claimed that all axons end in gray matter (seriously challenging the network, or reticular, idea) and that these endings are arranged *closely* around other cells. While this insight seems slight now, in 1887 it was a revelation. Cajal, without a network bias, perceived that "it is a regrettable truth that almost no one can be entirely free from the traditions and fashions of his day. Despite his admirable originality Golgi was badly influenced by Gerlach's doctrine of diffuse interstitial nets."

In a curious aside, a Norwegian student of Golgi's was stumbling toward the same insights that Cajal was chasing at about the same time. Although he never really got to the synapse, the dashing Fridtjof Nansen, who later won the Nobel Peace Prize, certainly had doubts about the network theory. In his 1887 dissertation, Nansen offered support for Golgi's classification of nerve cells and for the trophic function of dendrites, but in a sharp departure from his teacher's conviction, he found no connections between cell processes. Though he agreed with most of Golgi's histological ideas, Nansen explicitly rejected the network theory of Gerlach. The Norwegian was a vigorous young man with many interests and probably didn't realize the full significance of this work, even after it was published. Regardless, at the age of twenty-seven he abandoned neurohistology, skied across the Greenland ice cap with five others, and soon after sailed to the Arctic. In Norway he was revered, and he became internationally famous as an explorer, statesman, and humanitarian. Nansen's single inspired paper on the nervous system, and the more extensive work of August Forel and Wilhelm His, together were beginning to imply that axons ended.

If axons really halted close to another cell but left a space between them, then Cajal could focus precisely on this ending of one cell and the beginning of the next. His mission was to find examples of such anatomy. Demonstrating the conformation of the space between cells, he argued, would prove that neurons are the basic element of the nervous system. In only a few years of searching, Cajal had concluded that the nervous system is composed of units: a cell body with dendrites that receive incoming information and propel the nervous impulse down a single departing axon to stimulate the next cell. Finally, he deduced that since there is no continuity of substance between neurons, the electrical impulse between cells must be transmitted by an unknown chemical induction effect. While the precise chemical nature of this induction effect (now called neurotransmission) could not be known by light microscopy, Cajal's ideas incorporated all that is now felt to be fundamentally true about the basic structure and function of the nervous system. At that time, however, the matter was both imperfectly understood and hotly contested.

After he developed the double impregnation stain, he

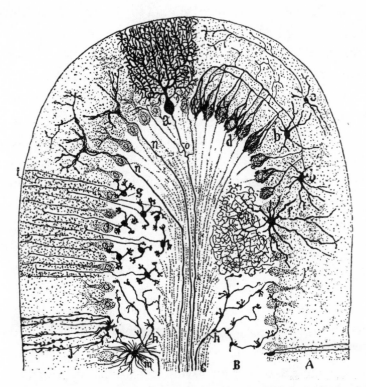

Transverse section through a cerebellar convultion (or fold) in the brain of a mammal. The vocabulary used here (molecular layer, granular layer, stellate cells, etc.) is simply description, names Cajal adapted to identify what he saw. A: molecular layer; B: granular layer; a: Purkinje cell with its dendrites spread out in the plane of section; b. small stellate cells of the molecular layer (basket cells); d: descending terminal axons from stellate cells embracing the cells of Purkinje; n: climbing fibers with terminal axons spreading over the dendrites of Purkinje cells. *From Cajal,* Recollections of My Life.

started again to examine the embryonic cerebellum of birds and mammals. This work eventually produced some of his most convincing images of neuronal connections. These drawings of Cajal's are magical and at the time they were first seen must have caused the same excitement created only eighty years later when Neil Armstrong walked on the moon. Cajal followed the branching axons of small stellate cells (b in figure) to their terminations as baskets (d in figure) around the large bodies of the cells (a in figure) that Purkinje had already discovered, and which are named for him. As can be seen in Cajal's drawing, the axons of the stellate cells actually do form a sort of basket surrounding the large Purkinje cell body. These rich endings provide thousands of inputs to each Purkinje cell, a requirement that ensures smooth coordination of movements. The anatomy of the axonal ending conforms to receivers on dendrites and the cell body, the way the fimbria (or fingers) of Fallopian tubes approach and lure the ovary, their cilia summoning the egg. Nearby he found the odd mossy fibers with strange tuberous appendages in the granular layer of cerebellum articulating exactly with "claw-like arborizations of the granule cells." He found the tiny spines on dendrites of Purkinje cells that increase their surface area and collect incoming messages. He recognized the parallel fibers of the granule cells (a constant in vertebrates from fish to man, as would later be shown by the anatomist Pedro Ramón y Cajal, who had followed his older brother's example after all) and the climbing fibers that connect cell bodies in the brain stem to the dendrites of Purkinje cells. Repeatedly Cajal found examples of axonal endings delicately conforming to the dendrites of the next cell. He found so much that it was difficult to record it all. Still he managed to draw what he had found and to write paper after paradigm-altering paper.

Few people read them. It was not that scientists in the rest of the world had gone into hibernation; they simply didn't know what Cajal was doing. They were busy at their own research and convinced of their own insights. Spanish science was either unknown to them or felt to be insignificant.

Novel approaches to brain research employing a variety of techniques for unraveling neuroanatomy and physiology occupied scientists in dozens of laboratories scattered around Europe. In Russia, Vladimir Betz developed an enormously complicated procedure to fix, stain, and section brains, writing papers based on his own examination of more than five thousand histological preparations. After he published this opus in an 1874 German journal, clearly identifying five cortical cell layers, as opposed to the mere lines visible to Gennari and others in fresh, unstained brains, he found evidence for structural specialization that he believed correlated with specific functions. These giant cells of Betz proved to be those cortical neurons specifically related to movements on the body's opposite side. This organization of "Betz cells" not only implied a center, or focus, of cells in the brain that governed movement but also a complex anatomy of differing cells within the center. In describing this himself, the Russian said, "Every part of the cortex differs structurally from other parts of the brain [and] this structural differentiation. . . . is the expression of the localized functions in the cortex."

Confirming Betz's anatomical discoveries, two young physiologists in Germany stimulated a dog's cortex with electricity and produced movement of the opposite side of the animal's body. There could be no doubt: Certain locations in the brain clearly presided over specific actions.

Studying injured patients in France, the neurologist Paul Broca had already localized motor speech, the physical ability to produce words, in the left inferior frontal gyrus (the lowermost part of the frontal lobe, just at its junction with the parietal and temporal lobes). A stroke or a wound in this small bit of brain—often no larger than a dime—left patients able to understand language but with words forever fastened to the tips of their tongues. A few years later, Carl Wernicke placed receptive speech, the understanding of language, nearby, in the posterior superior temporal gyrus (where the temporal lobe abuts the parietal). Patients wounded in this part of their brains could talk but spoke a nonsense language that we now call "word salad." When he understood receptive aphasia, or the inability to make speech, as distinct from Broca's expressive aphasia, Wernicke also began to unravel the connecting (or associative) functions of cerebral organization. While for preindustrial anatomists there may have been particular brain centers in charge of every function and feeling, Wernicke's discovery showed that they could not be completely autonomous but required association neurons to connect them. Sensations might enter the mind in certain places, but then they had to be recognized, related to memories, formed into ideas, and ultimately understood in a way that made possible their expression in language. Anne Harrington, professor of the history of science at Harvard, has said that "the new cartography of mind and brain that emerged was consequently a road map rather than a psychological geography." While there might have been centers for hearing and seeing, feeling and moving in the eighteenth-century villagelike models of brain (seen as autonomous clusters of artisans and craftsmen), in the nineteenth-century factory constructs of cerebral organization, association neurons, like relay runners passing the baton, were required for these regions to communicate with each other.

In Vienna, Theodor Meynert was drawing the same five cortical cell layers Betz had seen. Meynert's drawings, published in 1885, are a simple and accurate representation of these layers, but cartoonlike in comparison to the perfect renderings Cajal produced a few years later. In addition to microscopic research at the cellular level, more sensitive galvanometers soon allowed physiologists to record action potentials (discharge of neurons) and conduction times accurately. When they encountered unexpected delays in the transmission rates of electrical impulses, they began to suspect that barriers to current flow existed along the route. These delays could even be measured precisely at about two thousandths of a second. Maybe the slowdown was due to gaps.

The influential embryologist Wilhelm His, who believed in only what he could see, took a different approach to studying the developing cells of the nervous system, but a uniquely profitable one. In a paper published in 1886, he

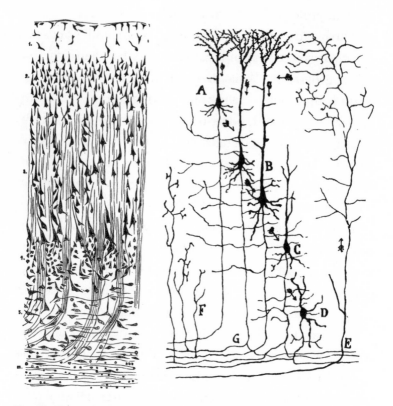

Theodor Meynert's 1885 representational drawing of the then known five num-
bered layers of cerebral cortex (left), compared to Cajal's detailed drawing of the
same layers published seven years later (right). Cajal later described the now
accepted sixth layer. *From Meynert, "Psychiatry: A Clinical Treatise on Diseases of
the Fore-Brain Based Upon a Study of Its Structure, Function and Nutrition," and
Ramón y Cajal, "El neuvo concepto de la histologia de los centros nerviosos. III—
Corteza gris del cerebro."*

showed that developing axons appear before dendrites and end freely near both motor fibers and peripheral sensory receptors. Embryonic axons ended! He could find no evidence to support the reticular theory and claimed that neurons therefore might be solitary units. August Forel, another Swiss scientist, who was examining mature brain cells in a different lab about that time, came to the same conclusion. Although many of these histologists were diligently using Golgi's silver-staining technique by the 1880s, few found support for his networks. However, Golgi himself still did.

Sequestered in Spain, Cajal toiled alone without recognition or acknowledgment. Although he had some access to the journal articles produced by these other laboratories, none were translated into Spanish, and so the university library saw no reason to buy them. Out of his always meager salary, now strained further by the growing family, Cajal subscribed to as many foreign journals as he could afford. When they finally did reach Barcelona, weeks or sometimes months after they came out, he could read the few in French, but the other languages of Europe were beyond him. Sometimes, though, he could gather meaning from the figures and from the cognates to Spanish that he recognized, which tend to be more frequent in scientific writing than in other prose. Even so, Cajal later wrote that 1888 was his greatest year, and that he dedicated himself "to work, no longer merely with earnestness, but with fury." He might have remained a respected professor at a provincial university in a country with little science, but silver stains and embryos opened the window for his fertile imagination. The fruit of this heroic amount of work, he remembered years later, soon would allow him to claim that "the trench of science had one more recognized digger."

SEVEN *Climbing Fibers and Basket Endings*

CAJAL HAD PLENTY of reasons to consider himself an innovator in 1888, but he was more than hopeful if he thought that anyone else might. When he published his original findings now in the several Spanish journals—ideas that penetrated a mystery debated over thousands of years—the world was immune. The well-known histologists in Europe, England, and even the few in America were unaware of work being done in Spain because no one bothered to translate it. Even when he managed to have a paper or two accepted in German journals, their readers had never heard of him and so paid little attention to what he wrote. While he was wounded by this disregard, he was not paralyzed. He had been toughened by his father, village life, and the boys of Ayerbe.

It is true that his seclusion in Spain certainly cost him the collaboration of talented colleagues and publications, but his isolation also protected him from some of the controversies heating up at that moment. One of these arguments was an attack on the Golgi method itself.

For years various schools had debated the existence of a

lymphatic system in the brain. Lymphatics, that pipeline similar to the arteries and veins, but interrupted by nodes, exists throughout most of the body as a sort of secondary treatment plant. This unique system specializes in returning lymph and cellular debris via its own duct at the root of the neck into deep, large veins. No less an authority than Wilhelm His had argued for the existence of lymphatic channels in the central nervous system, and a variety of others believed that the spaces around cells and blood vessels in the brain (which the young Golgi had examined) were really lymphatics. In Germany, several histologists, claiming to admire the Golgi method, simultaneously proclaimed that the silver was deposited not in nerve cells at all but in these phantom brain lymphatics. Golgi was livid at the attacks on his work, and he was right. It has now long been proven that lymphatic channels do not exist in the brain. What the Germans had actually seen were dendritic spines. Later Cajal saw them as well, but he knew what they were.

Golgi had seen these spines, too, but because he was by then convinced that dendrites were not part of neural transmission, he hurried by them. Rather he continued to believe that dendrites had a trophic (or supportive) function, a role he refused to assign to glial cells even when this truth became generally accepted. Because he never allowed himself to think that dendrites might be the collecting agency in neural transmission (that is, they receive the stimuli of neighboring neurons) and further thought that nerve cells were capable of conducting in all directions, Golgi's fundamental theory lacked cohesion. Still he was able to state with great certainty that

> the so-called protoplasmic process [dendrites] under no circumstances, either directly or indirectly, gives rise to nerve fibers and maintains independence from them . . . [and that] the diffuse neural net is distributed throughout all layers of gray matter of the central nervous organs and represents the system connecting sensory and motor cells.

When he began to investigate fine structure with silver staining, he was not alone in these views. But as more and more data arrived from a variety of laboratories in different disciplines, belief in the reticulum theory began to weaken. Golgi's own discovery of axonal branching had so validated the reticulum for him—a position solidified by his misunderstanding the receiving function of dendrites—that he began to discount the findings of others. As the historian Thomas Laqueur wrote about the reassuring blurred images made by Golgi's contemporary, the fin de siècle Viennese photographer Emil Mayer, "How much of what we see is what we hope to see in the world around us." Wherever he looked on his slides, Golgi found branches on axons that confirmed for him the network of the nervous system. Their processes formed for him "a system of ramifications. . . . that by subdividing, intertwining, and twisting in the most complex and bizarre manner, can, in fact, give rise to a uniform web, reticulum-like." He was intellectually prepared to find networks because he so fervently wished to believe in the holistic nature of cortical function—an idea already firmly denied by Virchow, had Golgi but reread the text that had started him on his journey of experimentation twenty years earlier.

Golgi's biographer Paolo Mazzarello speculates about the influences of Pierre Flourens (who helped to expose the phony science of phrenology) on Golgi's thinking. Early in the nineteenth century, Flourens had conducted ablation experiments in birds, killing specific layers of their brains. He had concluded that the cerebral cortex functions as a unitary organ of intelligence, without areas of specialization. Such an interpretation is particularly odd in that the same Flourens was also one of the first to identify several highly specialized regions of the spinal cord and brain stem, including the respiratory center in pigeons. While he associated these more ancient anatomical regions with specific functions, like many pre-Darwinian scientists, he philosophically exempted thought. He assigned to man a unified brain structure directed by the soul.

Flourens stumbled into this holistic maze because he didn't realize that, while the technique of sequential ablation successfully killed layers of cortex and so naturally enough altered both behavior and the senses, along with the layers he was killing the centers. This is a little like gathering tulips by pulling up the bulbs, and then assuming that the missing flower was what caused the plant not to reappear the following year. Neither did he have enough understanding of the enormous difference between the almost perfectly smooth brains of birds and the convoluted cortex found in higher mammals. Thus when Flourens speculated that thought arose diffusely from the entire cortex acting as a unitary organ of intelligence, the Frenchman sided with his country-man Descartes, separating mind from body and in the same stroke creating an expression of an overseer soul distinct from matter.

Though Golgi was too much of a positivist to go for such metaphysical ideas, Mazzarello does say that "there is no way of knowing how much Flourens influenced Golgi's initial holistic choices, but Golgi cited him in 1882 in his essay on cerebral localization, and continued to cite him and declare his admiration for him again and again in the following years." Golgi himself wrote that his thinking had moved closer to that of Flourens as he distanced himself both from the neuron theory and from cerebral localization, just when others were identifying more of these specialized areas of neural function.

Golgi was dogmatic when he lectured or wrote in support of the network idea, and often curt in his defense of the subject. As more laboratories published results supporting neurons and the theory gathered advocates, the Italian seemed to take many of the claims as a personal affront. Part of his antagonism focused on the issue of collateral fibers, which Golgi—probably correctly—always claimed to have described first. These fibers had been lost in the volumes of work he had produced after discovering the black reaction, and it soon became clear that he had misinterpreted their meaning, so his priority was often overlooked. In an 1890 paper complaining about this slight, Golgi wrote that in many recent publications on the topic, his results "were not even cited; and these same works, which in addition contained

only a fraction of the findings I published a long time ago,
were presented as if entirely new!" While the reticulum was a
logical framework for the discoveries in neurohistology during the middle part of the nineteenth century and nicely
explained the role that axons play in somehow connecting
cells together, it was losing adherents by the time Cajal began
to publish his more accurate work, for Cajal was modifying
not only Golgi's staining method but also his theories of neural conduction.

Many of the most influential anatomists of the period,
including Waldeyer, Retzius, and most notably Kölliker, had,
like Golgi, begun their investigations into the fine structure
of the nervous system convinced the reticulum existed. It
was a convenient explanation for what they couldn't see,
since they were looking then at the very limits of resolution of
their microscopes and using stains inadequate for the task.
As evidence accumulated throughout the 1880s, from discoveries in physiology, neurology, and even primitive neurosurgery, more and more scientists enthusiastically joined
histologists in jumping to the neuronal camp. Golgi's own
medical school teacher Panizza had localized vision to the
occipital lobes and (even before Broca) found speech in the
left hemisphere. By the time he began to produce his own
original research with the silver stain, a commitment to the
network theory obstructed Golgi's perspective on how the
brain was organized. Sidetracked by confusion over the function of dendrites and the meaning of terminal axonal
branching, he was never able to broaden his views. It is also
true that he was curious about many subjects, and his inter-

ests were shifting away from the nervous system just as the matter was being resolved.

The proud and perhaps emotionally isolated Golgi had too much invested in what he had already furiously defended to change his mind, so he simply summarized what he thought about the diffuse neural net and gave up writing on the subject. Reserved and always somewhat vulnerable if not overtly insecure, he could not admit that his observations might have been inaccurate or that he had made a mistake. Having put most of his vast technical effort into neurohistology, as Cajal redefined the discipline Golgi abandoned the field.

Medically an eclectic thinker, Golgi went on to produce influential basic research on kidney disease, intestinal parasites, malaria, rabies, and anthrax. But just as he had moved away from the nervous system, eventually he moved away from science itself. After he became rector of the University of Pavia in 1893 (and in a shrewd political move won a position on the Pavia City Council at the same time), his students began taking over as heads of major departments in histology, anatomy, and pathology throughout Italy. Golgi's successful career as rector eventually elevated him to a seat in the Italian Senate. Out of respect for their renowned teacher, his apprentices kept the network theory alive in major Italian universities long after it had been abandoned elsewhere.

We might understand Golgi best through the eyes of his orphaned niece, Carolina, who lived with her aunt and uncle after her own father's early death. For the childless Camillo and Lina, she became a beloved daughter. Golgi's emotional life seems, in fact, to have been confined to these two women. Carolina wrote that "to find in the life of my uncle events or episodes outside his life as a scholar, is not easy. He lived with his thoughts, relieved of all the daily necessities of life by his companion, for whom he was like a divinity . . . My uncle was a person of few words." After he left the bench of science, he used those few words to defend an idea by then in full retreat.

In Barcelona, Cajal heard little of this dispute. Bolstered by what did reach him from Europe—most prominently the findings of His and Forel independently supporting the neuron theory—he continued to search his slides for the secrets he was now certain they would reveal. Growing frustrated with tiresome editorial delays at the staid *Catalonian Medical Gazette*, he founded his own *Quarterly Review of Normal and Pathological Histology* so that he could publish more quickly. By 1888, he too had given up thinking about the nervous system solely in terms of anatomical description and was trying to divine how the wiring of the entire system actually worked. But just as Golgi retreated into politics, Cajal's chrysalis burst open to reveal one of the world's original neurobiologists, an integrator of disciplines.

Cajal wandered through his lilliputian world, carefully drawing all that he saw and also imagining the ways these tiny telegraph lines might send and receive messages. After all, the telegraph didn't transmit information just because there was a wire. A switch had to be opened and closed. In short, he allowed himself to speculate about why neurons looked as they did and what function their structure could serve. Within his first year and a half in Barcelona, he understood the synapse.

Over the ten years following his first startling discoveries on the fine structure of the cerebellum, Cajal had figured out the fundamental structure and basic physiology of the entire mammalian nervous system. Once he had glimpsed stellate cells branching terminally and then forming basket endings around the Purkinje cell bodies, the net, or reticulum, unraveled. When he found the rising axons of the long climbing fibers (Cajal referred to them as "robust conductors") ascending from the pons in the brain stem, traversing the granular layer of the cerebellum without branching and offering themselves to the Purkinje cell dendrites—exactly mimicking their contours—the network was dead. He had seen that axons end, by this time in the cerebellum, and later in the retina and the olfactory bulb. Those endings invariably conformed to the shape of the dendrites of the next cell. There was no fibril of connection. There was a gap.

Recollecting this moment much later, Cajal wrote in a typically understated style, "This fortunate discovery, one of the most beautiful which fate vouchsafed to me in that fertile epoch, formed the final proof of the transmission of *nerve impulses by contact* [synapse]." The words that Cajal wrote to mean stimulation of the next cell were the Spanish noun *contacto* or the verb *contactar*. Cajal himself was consistent in using these terms, and so were his English translators. While the Spanish may be translated as the English word *contact*, in various usages it also means to "be in touch with." Throughout the nineteenth century, all the terms being used to describe neuronal cells and fibers evolved along with the science, and clearly Cajal meant to use *contact* to differentiate between a continuous network of fibers and single units.

Even in his autobiography, recalling many wonderful details of his life and work, nowhere does Cajal ever tell us exactly how he was able to see these areas of contact and to formulate what all the others had missed. Perhaps it is ineffable, the quality of mind allowed to the very few. He does say that in examining the ideas of others there may be found "the intellectual lucidity, the solid culture, the technical difficulties, and sometimes the brilliant finding of genius; but with these appear also the prejudices, carelessnesses, and equivocations of the man of science. Once they are discovered, these little mistakes are very useful." He also writes in many places that most great discoveries "have been found as a result of chance." While there is no doubt that improved microscopes and the Golgi black reaction came at the right time for Cajal, and that the theory of the neuron had reached its moment, this man's artistic nature and his isolation were advantages difficult for him to appreciate at the time. It was his good fortune not to bear Golgi's burden of defending too much popular theory, nor was he bombarded with information, either correct or incorrect, from others in his field. Alone in Barcelona, he was free to imagine what the nervous system looked like in three dimensions and to daydream, as an artist might, about how neurons make their subtle reports to one another.

By 1888, Cajal had begun to refute the reticularists, not with ideas but with data. About this time, he saw and correctly interpreted what are now called synaptic boutons, a slight swelling in the termination of axons where they conform to the dendrites of the neighboring neuron. These swellings had been seen and at least partially described by others, including Deiters and Golgi, but then disregarded. In fact, as late as 1891 Golgi wrote describing filaments that "terminate with small bulges or slightly ill-defined swellings; such short and extremely thin filaments, ending in small bulges, can be seen in great number, when one follows the course of the finer fibrils; for now, I am unable to make any speculation on their significance."

Integrating boutons with his other cerebellar observations, and soon his findings in the retina, Cajal had proven to himself that he couldn't find networks. But in science, negative evidence is weak. However, he was confident that he had also discovered important microscopic anatomy of the climbing fibers and basket endings, demonstrating for the first time that axons terminate at swellings that conform perfectly to the outline of the next neuron's waiting dendrites. This reaching out of dendritic spines (thorns, he sometimes called them) was further proof that this surface area was increased to "receive and propagate the nervous impulse," implying that dendrites ended freely instead of connecting to other neurons—the mistaken position Gerlach had introduced by claiming that protoplasmic processes met in a network.

Taken together, this information drew a picture of unconnected individual cells. What remained to be shown was the other side of the articulation—today called the postsynaptic membrane—on the cells and dendrites, thereby proving the existence of a gap, across which leaped a still-mysterious electrical "contact."

To find positive evidence for the synaptic space on the other side of the gap, the dendritic side, Cajal turned first to the highly specialized visual system. He found that the receptor cells—rods and cones—end freely in a densely interconnected arrangement of fibers known as the outer

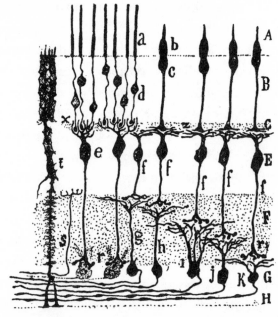

Mammalian retina. A: layer of rods and cones; B: outer granule cell layer; C: outer plexiform layer; E: bipolar cell layer; a:
rod; b: cone; c: cone cell body; d: rod cell body; e: bipolar
cell with dendrites above (x) surrounding the rod axons' terminal spheres. The importance of this figure is that, in the
outer plexiform layer, axons of rods and cones conform to
the shape of the dendrites of the underlying bipolar cells.
From Cajal, Recollections of My Life.

plexiform layer of the retina. The cell body of the cone, just under the cone itself, gives rise to a single straight descending axon that ends in this outer layer as a few filaments sprouting off horizontally. The rod cell body is bipolar, however, meaning that it has a rising thin fiber ending as the rod on the retinal surface (analogous to a dendrite that receives the stimulation of light), while the descending fiber terminates in the same plexiform layer as the cone's axon. Since the body of the cell is in between these two fibers, it is called bipolar. The descending fiber of the rod ends as a tiny sphere with no branches, but it is still an axon. Not only do they *end*, however, but the axons of both the rods and cones also conform exactly in shape to the dendrites of large bipolar cells in the next layer down. Eureka! This is a circumstance, Cajal notes, heretical to the reticularists. Their network theory required that such important cells as rods and cones—the basis for the vital sense of vision in mammals—could not be so isolated but would have had to fit directly into a network of innumerable fibers occupying the outer plexiform layer.

But there was no such fit. Instead Cajal held that the impulses from the rods and cones must rather be "gathered by certain bipolar cells and conveyed individually to certain nerve centers." Here, as far as it could be outlined before electron microscopy, is the exact image of a postsynaptic membrane. Where Golgi had seen the tunnels of an anthill, Cajal saw the ants.

He still didn't quite see individual ants, though, but rather columns of them. He knew from trying to give anatomy lec-

tures to medical students that the existing knowledge of
spinal-cord architecture was rudimentary. How did messages
arrive in the cord through the sensory nerves and make their
information known? Where did it go? The sensory ganglion
was understood, with its strange bipolar cells, and the sensory root fibers could be found entering the cord in what was
already called the dorsal horn. There the fibers vanished into
jumbled speculation, reappearing when the large motor
neuron cell bodies extended axons that soon coalesced (at
the ventral horn of spinal cord gray matter) and became
the exiting motor roots headed out to muscles. The network
theory had helped explain the confusion of the white matter
separating dorsal-horn stimulation from ventral-horn
effect.

Still anatomists searched in the spinal-cord white matter
for other clues beyond its basic organization. Researchers
had studied the spinal cords of patients who had died after
accidents left them paralyzed. Others made experimental
lesions in the cords of animals, causing weakness, numb-
ness, or imbalance in a specific limb or limbs. They could
find collections of fibers with identical functions ascending
and descending as tracts in the white matter, but what these
fibers meant to the cells in the brain and spinal cord was a
mystery. The puzzle was made even more complex by the
perverse habit some fibers had of appearing to cross from
one side of the spinal cord to the other before they formed
the tracts. All this data had been collected from studies on
adult humans or animals, using stains that revealed everything except what they were trying to find.

Histologists up to this time had been limited by nonsilver stains that exposed the cell bodies and myelinated segments of axons in spinal-cord white matter but failed to reveal the terminal, unmyelinated segments—exactly what the scientists were trying to see. Cajal had saved a notebook containing his old drawings of the cord, which he had prepared when studying for his doctoral examinations. Taken from neurological textbooks, they showed three "perfectly irreconcilable schemes" of how the connections worked. Using his modifications of the Golgi method and examining the spinal cords of embryos to minimize the inhibiting effect of myelin, Cajal set out to sort through this disarray.

In launching this massive undertaking, Cajal already had the blueprint he had made in the cerebellum. Whereas the reticularists had drawn interconnections like the tunnels constructed inside anthills to represent the white matter of the cord, Cajal had his eye on individual ants. He felt certain that he could find individual connections between the incoming sensory fibers and neurons in the cord, an idea he by this time called the "*law of pericellular contact.*" He had embryos with fibers still largely unmyelinated. He had silver to stain the unmyelinated terminal segment of the axon. He had a microscope just good enough, and he had his indestructible determination and concentration.

His meticulous drawings picture separate sensory fibers entering the cord, breaking into long and short collateral branches and then, almost with desire, being "applied intimately to the contours of the bodies and dendrites of the neurons." At last he had before him the commissures where

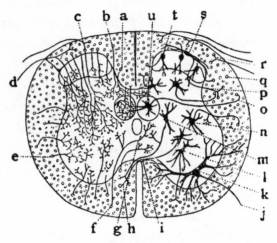

Diagram of a cross section through the spinal cord as seen
by Cajal using his modifications to the Golgi silver-staining
method. Note the complex details of neuronal connections
Cajal saw in gray matter between the sensory neuron axon
entering and the motor neuron axon leaving the cord. These
collateral connections are essentially all the letters except d
and r: sensory axons; j: motor axon; k: motor neuron cell
body. *From Cajal,* Recollections of My Life.

bundles of fibers crossed, the columns of ascending and
descending tracts in white matter, and the true terminal
arrangement of sensory fibers on intermediate and motor
cells.

At the same time he permanently disproved theories that
fibers controlling movements all cross in the cord (the vast
majority actually cross in the brain stem) and that sensory
neurons are a continuous web of fibers. He had found the
individual neurons and their synaptic meeting places.

There was no doubt in Cajal's mind that he now was close

to understanding the basic form and function of the neuron, an idea more serviceable than the reticulum, or network. Cells were individual units communicating with each other across a gap, and this gap, the synapse, was proven by the precise way the endings of axons and the beginnings of dendrites suited each other.

With this understanding came a grasp of the fundamental organization of the tracts in the spinal cord and the specialized nature of certain regions of the brain. Neurons must be bipolar, he was certain, and could conduct in only one direction. Now all he had to do was convince the rest of the world.

EIGHT

Sincere Congratulations Burst Forth

As THE NINETEENTH CENTURY moved toward its final decade, the best science remained German. True, the Britons Lyell, Darwin, and Huxley had already written brilliantly on geologic time, natural selection, and evolution: Their ideas had taken firm root in England and were spreading. Even so, for most of Europe and the rest of the world, study in the natural sciences followed the lead established in German universities.

A Swiss researcher imported to the United States in the middle of the century, Louis Agassiz—an expert on the Ice Age, a fine collector, and a better showman—sullied his reputation after he sided with the phrenologists (and in the process with slaveholders) by championing a flawed, if not faked, analysis of cranial capacity in his public lectures. The pragmatic school of Oliver Wendell Holmes, John Dewey, and William James was gathering influence, and the ensuing movements in social science and politics generated novel theories and reforms. But still bench science in America remained inconsequential. By and large German scientists set the agenda for what to investigate, and German editors determined what was fit to publish. Thereafter they argued among themselves about these reports, occasionally allowing the French, Italians, Scandinavians, and Swiss to comment. So it was to the Germans that Cajal turned for approval.

Cajal was genuinely perplexed when his publications, which he knew contained revolutionary proof of the not-yet-named synapse, attracted the attention of only a few in the European scientific community. When his work was noticed at all, it was usually to acknowledge the support it lent to Golgi's methods and contributions. Others dismissed his ideas as foolish and beneath consideration. With crippling logic, one Swiss anatomist complained that Cajal's findings about the spinal cord must be wrong because so many others had, for more than fifty years, failed to make similar observations! These reactions were certainly not what Cajal had anticipated when in excited hope he had sent off complimentary copies of his *Quarterly Review*. Searching German, French, and Italian publications for references to his work, he found none and concluded, optimistically, that few histologists read Spanish or bothered with translations. While that was true, it wasn't the only problem. Like many revolutionary ideas in science, including the significance of the black reaction to begin with, the real meaning of his work was at first buried under a pile of prejudices.

Though a gregarious man who would later attract an international following, Cajal wasn't a self-promoter. He agonized over how to make his discoveries known and finally settled on a two-stage strategy. Beginning in early 1889, he trans-

lated his own papers into French and submitted them to the
best German journals. The German editors considered
French a developed-enough scientific language to be taken
seriously. At least this effort surmounted the chauvinism of
language, but it was not enough. The real power in Cajal's
arguments had to be seen in the slides themselves. Anyone
can draw what he believes he ought to see, and plenty of histologists had vivid imaginations. He had to *show* them the
proof.

His second step, therefore, was to apply for membership
in the German Anatomical Society. In an era without easy,
prompt communication among researchers, gatherings of
professional societies were where the newest research made
its way forward. Mail was slow, and nobody was paying much
attention to anything Cajal wrote in letters or his journal any-
way. The annual gatherings of international scientific organizations—now, because of faster communications, pro-
gressively more tradition than necessity—were at that time
essential for the fermentation of ideas. Scientists eagerly
looked forward to these meetings so that they could present
their own research, get an idea of what others were doing in
their labs, and perhaps even poach on someone's intellectual
property. Furthermore, the meetings were held in the most
notable hotels and universities in the grandest cities of
Europe. Beautifully dressed men rode through town in elegant coaches to the scientific venues, returning late to their
hotels. They entertained one another in the drawing rooms
and dining rooms of these hotels, and could be found
together drinking in cafes or sightseeing. Distinguished pro-

fessors made formal presentations on large topics, and younger investigators were given an opportunity to present their own less exalted research. Manufacturers brought the latest technical advances: new microscopes, sources of light, dissecting equipment, better glass slides, and reagents. Rooms were set aside for demonstrations of the new histological findings, which could be viewed through microscopes on the original slides.

In October of 1889, when the Anatomical Society was to meet, Cajal scraped together his savings, as his institution had no money to pay for the trip, and boarded the train for Berlin. This was really one of his first opportunities to visit other laboratories, so he didn't hurry his journey and stopped at several universities along the route to Germany. He visited Lyons and Geneva, and stayed for a few days in Frankfurt-am-Main, a town then without its own university, but still a venue for first-class science. There he met Carl Weigert, discoverer of the staining technique named for him, and Ludwig Edinger, already a celebrated neuroanatomist (and a fine artist) who had recently described previously unknown tracts and nuclei of the spinal cord and brain stem.

Weigert had been the beneficiary of William Perkin's 1856 accidental discovery of industrial chemistry. As an eighteen-year-old student in London, Perkin had derived the first aniline dye from coal tar. The mauve color he produced quickly became important to British dyers, but his discovery also opened the door for production of synthetic dyes suitable for staining cells. Weigert stained bacteria in tissue with methyl violet in 1875 and encouraged his cousin Paul Ehrlich in his own studies showing that aniline stains not only colored tissue, but actually formed chemical reactions with nucleic and amino acids. In his Berlin laboratory seven years later, Robert Koch used Ehrlich's methods to prove that the rod-shapped bacillus he had cultured and then stained with methylene blue caused TB.

Berlin in the fall of 1889 was a great cosmopolitan metropolis, home to an important university and the hub of European science, but it was also a city filled with apprehension. Chancellor Otto von Bismarck, the statesman who had unified Prussia with thirty-eight other German states, was at first annoyed and then alarmed by the behavior of Kaiser Wilhelm II. After Wilhelm I died following a long reign, his already terminally ill son Fredrich III momentarily succeeded to the throne but governed for only three months before he too died of cancer. As the body lay in state, his own son, the new kaiser, Wilhelm II, immediately demonstrated his love of martial authority by surrounding the palace with soldiers. Ignoring his dowager grandmother, Queen Victoria's, disapproval, he failed to observe what she considered to be a suitable mourning period before he sailed with his fleet up the Baltic to Russia. The kaiser's fascination with his navy, and especially the dreadnoughts, eventually proved disastrous not only for Germany but also for the world.

His cousin the tsar of Russia found Wilhelm II "a rascally young fop who throws his weight around, thinks too much of himself, and fancies others worship him." The new ruler's adolescent foolishness concealed considerable ambition, later ignited by even larger insecurities due, in part at least, to the brachial plexus palsy that had deformed his left arm at birth. The unhealthy combination of the young man's ambition, will, and anxious uncertainty would require Bismarck to craft a different relationship with Wilhelm II than he had had with his father and grandfather. But he didn't yet appreciate that the relationship would be quite so brief; already the twenty-nine-year-old kaiser was plotting the elder statesman's retirement.

While the chancellor had increased the size and status of the German army and seen to it that a comfortable working class and a rising middle class enjoyed rapid industrialization, he had also held tight reins on both. In his craving for military display and attention, Kaiser Wilhelm II at this time played an oddly liberal but disruptive role in the politics of the staid German society. Labor disputes spread, the Reichstag split on the issues, and soon Wilhelm barely spoke to his aged chancellor. This situation grew worse over the summer of 1889 when 170,000 Westphalian coal miners went on strike and Wilhelm insisted, against Bismarck's advice, on publicly siding with the miners.

The social unrest that raced through Germany was still simmering by the time scientists arrived for the Anatomical Congress that fall. While people in the streets of Berlin weren't worried about their safety, they were not contented. In addition, the city still buzzed over the scandal at Mayerling. The Austrian Archduke Rudolph had murdered his seventeen-year-old mistress in his lodge, then killed himself, thereby promoting his cousin, the unfortunate Franz Ferdinand, heir of the Hapsburg dynasty.

Completely focused on his science, Cajal either didn't notice or ignored this tense atmosphere and settled into his modest hotel near the University of Berlin. The next brisk early October morning, he walked to the opening session of the meeting, where he sensed surprise and even a certain chilliness from the men to whom he introduced himself. There were only a few hundred people in the world who understood the work being presented, most of them knew one another, and all were startled to find a Spaniard in Berlin at an important anatomy meeting. Who was this slight, balding, bearded man who spoke poorly accented French and presumed that he might contribute to an understanding of neuroanatomy? A Spaniard? In Germany!

During the formal presentation of papers, which still today is often a soporific moment, Cajal's mind wandered. He was impatient to move into the ornate demonstration hall, flooded with sunlight from the tall windows, and set up his slides. He unpacked his cherished Zeiss microscope, borrowed a couple of lesser models, and carefully mounted the slides under the oculars. In order to prove his points about the synapse, he had come armed with his best examples of structures he knew no one else had ever seen: beautiful slides accurately revealing individual cells communicating with each other, axon to dendrite. With an almost unendurable excitement he focused his microscopes on the baskets formed around Purkinje cells, rods and cone axons ending on bipolar cell dendrites, and perfectly outlined pathways, intermediate cells, and tracts connecting sensory input to motor output in the spinal cord.

He waited. Nothing happened.

The handful of people who had seen copies of his journal didn't know who Cajal was, and even these few mistrusted his science. The French were indifferent, the Italians argumentative, and the Germans contemptuous. In his hesitant French, Cajal attempted to gather those he hoped would examine his masterpieces, but most of them were more interested in showing off their own work and quickly wandered away. Finally, growing desperate, he warily approached Albert von Kölliker. Against the great man's protestations of too little time and the resistance of his obvious indifference, Cajal at last herded him to the far corner of the demonstration hall where his slides were on display.

Kölliker sat down. He was a man who attracted a crowd wherever he stopped, and soon a collection of both famous and ordinary anatomists surrounded him. He peered at Cajal's specimens. The others watched. Would he scoff, or simply say nothing and leave? For a long time the white-haired dean of European histology studied the slides, slowly realizing that before him, laid out in mesmerizing detail, were the solutions to dozens of intellectual problems.

He backed away from the microscopes and turned to Cajal with delight in his eyes. In French too rapid for Cajal to follow, he asked him for the technical specifics of his methods. As they stood in the huge room, surrounded by a growing crowd, Kölliker repeatedly returned to the microscopes, pointing to certain cells and demanding details. Why were these slides perfect and clear, when his own Golgi stains so often disappointed him? How was it done? What tissue did he use? How was it fixed and cut? How long did it remain in the silver solution? What were the concentrations of the various reagents? Could these contacts be shown in other parts of the brain, and in other species? What tricks had produced such miracles?

Even as he haltingly explained the particulars of his methods and Kölliker publicly and effusively congratulated him, Cajal recalled, "Finally, the prejudice against the humble Spanish anatomist vanished and sincere congratulations burst forth." Gustaf Retzius, His, Waldeyer—the greatest histologists of the era—crowded in next to Kölliker at the microscopes. When they finally looked up, they agreed with the father of German neurohistology that they had witnessed a scientific breakthrough.

As Kölliker left the hall, he asked Cajal to join him for dinner. They rode together in his carriage to a fine restaurant, avoiding the just-marketed Benz automobiles, where the abruptly celebrated Cajal was introduced to more great scientists. August Forel was there from Zürich, already a proponent of neurons based on his own work, and that of His. The brilliant, odd Theodor Meynert, who somehow inspired the work of both Forel and Freud in spite of a reputation as a poor teacher, had come from Vienna. Franz Nissl also attended the meeting on the way to take up his recent appointment in Frankfurt. As a medical student in 1885, Nissl had won the neurology prize for experiments with his innovative staining technique revealing new structure in the cytoplasm of neurons.

Now they were all eager to meet Cajal. Remembering that

meeting twenty years later, Retzius said, "Cajal's first studies had an electrifying effect upon those who were working in the same field . . . Albert von Kölliker and I were enchanted by the sight of the preparations which Cajal placed before us. Both he and I were converted and we started home again to begin working afresh with Golgi's method, which was not in great repute among other anatomists of that day."

Ending an evening that must have been emotionally overwhelming for Cajal—as well as linguistically exhausting—Kölliker promised to adopt his new methods immediately and to attempt to confirm his findings. As they parted that night, Kölliker told him, "I have discovered you and I want to spread my discovery in Germany."

Over the next several years, the studies pouring out of German laboratories duplicated and expanded on Cajal's work.

Cajal's return to Spain was a delicious and languorous expedition through many more towns and many more laboratories, to which word had traveled fast. Now he was acclaimed and welcomed. His curious nature extended not only to the research being done but also to the foreign styles of education and social structure in several of the cities where he stopped—Göttingen, Turin, Genoa, and Pavia. Visiting these famous old universities and museums delighted him, but he also relished every opportunity to display his Spanish patriotism. While visiting a home in Göttingen, he took special pleasure in pointing out to a new colleague that the authenticity of the prized Velasquez hanging in his library was, at the very best, doubtful.

If the fake painting offended his sense of order, Cajal also doubted the rigor of German and Italian universities, when he learned that they appointed their professors without competitive examinations. He was even more shocked by the individual liberties afforded not only to the teachers, but to the entire academic community. Eventually he realized that such systems relied on a deep acculturation and the committed goodwill of the faculty, toward both their students and one another. In Spain, a country of hierarchy and favoritism, he thought, a system like this would soon result in "a state of savagery."

Events soon demonstrated that, on the contrary, it was in Germany that the savage was concealed. Wilhelm II proved more misguided despot than rascal, and within six months of Cajal's glorious moment in Berlin he had forced Bismarck's resignation. This in itself might not have proven disastrous had the kaiser been prevented from consolidating his power over the German bureaucracy—policies and procedures he was ill suited to administer.

For the time being, German science remained secure; however, Wilhelm's continued buildup of the German military, and especially his construction of the dreadnought battleships, charted a course that would lead to the disruption of the balance of power in Europe. In 1890, the kaiser ordered a copy of *The Influence of Sea Power Upon History, 1660–1783* by the American naval historian Alfred Mahan placed aboard every German warship. In order even to move the monster dreadnoughts between the Baltic and the North Sea, the Kiel Canal had to be enlarged and deepened.

British politicians grew alarmed. The money and talent

illogically drained away from German universities and industries by these schemes eventually helped bankrupt their revered science and, along with the kaiser's ambition, ultimately encouraged the Great War

CAJAL WAS UNAWARE of these rumblings. When he passed through Pavia, he only lamented that he couldn't meet Golgi, who was in Rome being installed in the Italian Senate. In an August 1889 letter to the Italian, Cajal had proposed this visit but in his enthusiasm had also perhaps unwisely included the remark that "my preparations are so clear, so analytical, that all doubts concerning certain facts are absurd."

Coming from an unknown Spaniard, this slight to the network theory would surely have been unwelcome. Even at this moment early in what would become their great debate, both men may have foreseen the chasm opening between them. Nevertheless, Cajal's optimistically industrious and yet sanguine personality led him later to conclude that if they had met early in 1889 and discussed scientific ideas man-to-man, their quarrel would have been avoided. Cajal had enormous respect for Golgi and, given the chance, in a personal meeting would certainly have made that admiration known. But getting through to the man beneath the scientist would have been difficult, even then perhaps impossible. Cajal's mind was open; Golgi's was made up.

Cajal's return to Barcelona sent him into an intense period of new work. His experiences in Berlin authenticated his science and filled him with confidence and a renewed passion for observation. Fifteen hours a day at the microscope were now routine for him, and he wrote about what he saw there with a lyrical wonder and a clear understanding of the arbor that was the Purkinje cell, the flower gardens of gray matter, and the hidden islands and virginal forms waiting since creation to be found. Cajal was on the way to paradise, not realizing that the way led through Italy and a gate guarded by the man he respectfully called the "savant of Pavia."

He first needed to shore up his claim on the neuron. As a corollary to finding that axons conformed to the architecture of dendrites, Cajal's work strengthened support for what, in two 1891 review articles, the German histologist Waldeyer baptized "the neuron." Though Waldeyer did the naming, and clearly enunciated the neuronist's position in the same paper, it was Cajal who had provided most of the scientific discovery. These papers of Waldeyer's combined the work of Cajal, His, Forel, and others into proof of the neuron and synapse.

To strengthen his position, Cajal again began at the beginning, with embryos. Controversy had surrounded the issue of how neurons developed embryologically and how they ultimately found the way to their targets. With the reliability of German railroads, peripheral nerves always connect specific sites in the brain and spinal cord to specific sensory receptors and muscles. As the neuron theory predicted, and as the studies of both His and Forel had suggested, neurons mature from a precursor cell body (neuroblast) into an axon, growing dendrites along the way and finally finding a way to their synapse. This view met with opposition from those who, wishing to explain the process by which cell body and axon invariably found the proper target, supposed that after the first cell division, one neuronal nucleus stayed with the peripheral sense organ while another remained with the nervous system. These two nuclei, they argued, still connected by cytoplasm—a sort of bridge continuously under construction during development—then served as a template for the the rest of the nerve fibers to follow. Hensen, the champion of this idea, further complained that no one had actually ever seen a free nerve ending in an embryo.

Negative evidence, as we have seen before, is a weak argument. Cajal went to work to find the ending. Like His, he stained embryonic tissue with silver and saw that axons sprout early from their neuroblast cell bodies (in a chick embryo, by forty-eight hours of incubation), soon producing a swelling on the end. In scientific writing he called this swelling a growth cone, but he more commonly referred to it as a battering ram or club that mechanically forced its way to the target. Later in embryonic development, this growth cone branched terminally at the axonal ending, and at about the same time dendrites sprouted from the cell body. The battering rams on the ends of axons were somehow being *summoned*. The bridge theory collapsed.

These findings explained the development of peripheral nerves, and maybe even cranial nerves exiting the brain stem, but didn't offer much direction to axons wandering in the forest of the brain itself, or in the spinal cord. So Cajal returned to a jungle he knew well—the cerebellum. Again studying embryos and newborn animals, he discovered that his term *climbing fiber* was more than a metaphor. These

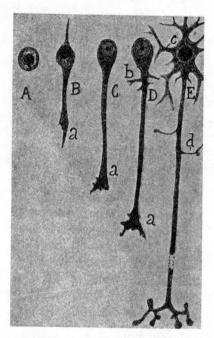

Maturation of neuroblast into mature neuron as seen by both His and Cajal. A:
embryonic nerve cell; B: bipolar phase; C:
stage of the neuroblast proper with (a)
growth cone; D: appearance of dendrites
(b); E: mature neuron with branched terminal axon replacing growth cone. *From
Cajal*, Recollections of My Life.

axons actually sought out Purkinje cell bodies and then
climbed out to approach their dendrites. He explained this
neuronal deportment by the theory of neurotropism, an idea
predicting a chemical beckoning that invariably lured axons
to their proper destinations.

Now not only Kölliker but scientists all over the world
rushed to embrace the neuron theory, and study after study

appeared confirming Cajal's work and Waldeyer's explana-
tions. Over the space of only a year, Cajal was raised from the
obscurity of an extra to a leading role on the stage of science.
Not only did he receive important prizes, titles, and honorary
degrees, but in events that must certainly have perplexed and
irritated Golgi, he was especially honored by the Italian
National Academy of Medicine and Pharmacy and elected a
Corresponding Member of both the Italian National Academy
of Medicine in Rome and the Academy of Medicine of Turin.

In all his writing, Cajal credited Golgi for the discovery of
silver staining, and for the enormous contributions he made
to neurohistology. But for Golgi this homage was never
enough. As the neuron theory gained strength, complaint
poured out of Pavia, arguing one trivial slight after another.
Responding to a letter from Kölliker about a paper published
in 1890, Golgi lamented, "I cannot believe that such a neglect
is completely justifiable, nor that I should continue to bear
it." His papers were not properly referenced, Golgi protested
to Kölliker; he had described axonal collaterals in embryonic
spinal cords first. At the same time he wrote that the structure of the gray matter of the spinal cord is so complex that
attempts to sort it out are "impossible and superfluous." Still,
he claimed priority. All these grievances recurred as he
defended his network theory against rapidly vanishing support. No one else could find Golgi's grove in the brain, shar-
ing like aspen a common system of roots.

Outside his circle of students, the Italian was isolated.
Work by Kölliker, Waldeyer, Retzius, His, Forel, and many
others all confirmed and expanded Cajal's findings: Neurons

are single cells consisting of one axon and many dendrites that never interconnect in the same cell; axons terminate in a variety of patterns, always associated with dendrites and cell bodies; electrical impulses travel in one direction, down axons and across the synapse to stimulate the cell body and dendrites of the next cell. The neuron theory was becoming law.

Over the next thirty years, Golgi proceeded to magnify and publicly demonstrate his own worst faults. Perhaps like the kaiser, he required and didn't have a Bismarck to advise him. The same habits of ambition and insecurity that, on a larger scale, allowed Wilhelm II to presume he could compel the world forced Golgi to cling to networks. Because he could not let go of that idea and appreciate the ascendancy of the neuron theory, the formative contributions he made to understanding the nervous system were to be obscured in scientific memory. Golgi is most often remembered, regrettably, for the size of his error.

But it was Golgi's research, especially his discovery of the black reaction and his detailed if imperfect observations on the fine structure of the nervous system, that had permitted Cajal to see the synapse. Cajal himself acknowledged this in a paper he wrote attempting to soothe the Italian over the issue of who first had reported the collateral fibers of axons in the embryonic spinal cord. This research was vital, of course, because it implied a basic understanding of how gray matter is organized in the cord—how sensory input reaches motor output—but it was far from the central issue.

How delicately Cajal applied himself to the entire controversy when he wrote, "I admire the work of Golgi and profess the greatest respect and utmost consideration for his scientific persona. We are indebted to him and to his seminal and path-breaking experiments for the precious method that allows us to discern the innermost structure of the nervous system with the clarity of a diagram; his great merits. . . . however, do not excuse him from acknowledging the modest merits of those who. . . . are honored to call themselves his disciples and followers."

NINE

The Prize

BY THE TIME the International Congress of Medicine met in Berlin in early August of 1890, most neurohistologists had already left the neuron and network controversy behind. It wasn't so much that the reticular theory had been wrong as that it was no longer a useful way to examine the problem. Scientific theories are little more than intellectual models devised to define a strategy for collecting data about the physical world. Scientists expect to formulate general rules when the data agree with the model. As long as the rules are not violated by what is observed, the theory is constructive. But when research data conflict with hypothesis, the assumptions are discarded and a new way of thinking arises, usually resulting in a better explanation—another theory. If this new hypothesis coheres and fits the data, it becomes a law, at least for a while. Thus science stumbles ahead.

Golgi attended the Congress of Medicine not to present evidence in support of his theories about the nervous system but to introduce new research on the life cycle of the protozoa responsible for malaria. In the halls and lobbies, though, he continued to complain about Cajal. The matter had become personal, not scientific, and the reserved and socially uneasy Golgi didn't have a character suited to resolving the conflict, either directly or indirectly. He simply got mad, and increasingly dogmatic. As the philosopher George Santayana said, "A theory is not an unemotional thing."

Cajal ultimately emerges from the chronicle of this dispute not simply as the superior scientist, but as the more mature and well-balanced person. After he had acknowledged Golgi's important work, he quietly withdrew from the quarrel and never returned to it, even when he might have been tempted. Now he was the purist, the most dogged observer of the nervous system. For several years he applied the majority of his effort to understanding the direction of electrical flow in neurons. This question was still a nonproblem for the remaining reticularists who, believing everything was connected to everything else in the nervous system, taught that the direction of conduction not only didn't matter, but also could change.

Writing that "the retina has always shown itself generous with me," Cajal initially turned to the visual system seeking hints about the direction of current flow in neurons. He found that the dendrites of sensory cells (first in the retina, later in the olfactory bulb and vestibular apparatus of the middle ear) are always *pointed outward* toward the external world, while the axon is always *headed inward*. This rule applies to the spinal cord, too, although the anatomy of the spinal ganglion cell is a little unusual in that the dendrite is very long. This peculiarity made some anatomists confuse it with an axon. In these sensory cells, the axon is the short fiber, but still it is directed inward from the cell body. This made sense to Cajal: The long dendrites collected sensory observations from skin and other organs and sent them to the spinal cord.

The situation is reversed in motor neurons. Internally directed dendrites receive information from inside the spinal cord and transmit their electrical impulse outward, along the axon and toward the world. Thus they instruct muscles to contract. Even then, Cajal recognized certain exceptions to these generalizations (for example, association neurons that connect entirely within the substance of the brain and cord), and more have been shown since. But the evidence available in 1891 allowed him to propose what he termed "*the theory of dynamic polarization.*" He concluded that neurons conduct only from dendrites to axon.

In attempting to broaden this idea, he was safe, he admitted, as long as he didn't venture too far into the convolutions of the mammalian brain itself. But the results weren't much better when he confined his research to the theoretically simpler brains of lower forms—reptiles and amphibians. Even the humble central ganglion that serves as a brain for insects proved impenetrable. He had, however, by then established the general idea of dynamic polarization, and physiologists soon expanded on his findings. While struggling in his studies of the cortex of these various species, Cajal did manage to show the phylogeny (evolutionary development) of increasing intricacy in the motor neuron cells of cortex from frog to lizard to rat. This research complemented his discovery of the ontogeny of the neuroblast in man (its development from embryo to adult) as the more basic cell matured to a neuron.

As the end of the century neared, he found it easy to publish his work and over a few years produced more than sixty papers. Somewhere along the line, he was astonished to learn his lectures to medical students had been collected and translated into German and French. When he was invited to speak at his old medical school in Zaragoza, he was met by a reception usually reserved for Spanish royalty. His lectures were a huge success, not only for the new information they revealed but also because this son of Zaragoza was now such a celebrity in foreign capitals as well as in Spain. International scientific reputation simply didn't come to Spaniards. For the local professors, these were tributes they considered beyond them and their students. Cajal's former teachers arranged a banquet for him, where they doted on him and offered lavish toasts in his honor. Even the redoubtable Don Justo had to admit that his son's artistic eye, which had so worried him twenty-five years earlier, was now producing sensational scientific portraits.

By April of 1892, less than four years after the Berlin meeting, Cajal was one of the most influential scientists in the world. In his own country he was rewarded with an appointment to the Chair in Anatomy at the University of Madrid. He was only forty years old.

Reading Cajal's one-hundred-year-old papers today, it is still staggering to calculate the labor that produced this mountain of perfect drawings.

About the time of his emergence, the always disorderly

oscillations in Spanish politics brought to power a dictatorial premier, soon followed by a liberal king, then a regent, and finally the vapid Alfonzo XIII, last and perhaps least of the Borbón rulers. Conservatives were returned to power in the elections of 1890, notable for voting irregularities at the polls and indifference by the electorate. The army was happy to play a role—as were Carlists who supported the monarchy, conservatives, socialists, Marxists, communists, anticommunists, liberals, and anarchists—dozens of factions all playing their part in a passionate confusion. Not even George Orwell could completely unravel these convoluted politics fifty years later when he wrote *Homage to Catalonia*, his firsthand account of the Spanish Civil War. The protectionist policies of the conservative government led the state to bankruptcy in 1892, just as Cajal and his family moved into a modest Madrid apartment and began trying to get used to life in the boisterous capital.

None of the turmoil disrupted Cajal, although he was sensitive to the potential seductions of urban life. He delighted in his work, as always, and enthusiastically returned to an intellectual cafe society with other scientists, philosophers, writers, and politicians. Associations of friends who meet regularly to debate are called *tertulia*, a Spanish word without a direct English translation but a common feature of Spanish intellectual life. The society that Cajal joined at the Café Suizo included one occasional member whom he especially admired, the liberal politician Segismundo Moret, later to be briefly the premier of Spain. While these men (there were no women in the group) were all patriotic, they also deplored Spain's ruthless 1895 attack and wholesale murder of the revitalized separatists in Cuba.

The United States had nurtured that rebellion, which led, at least indirectly, to the Spanish-American War three years later. The republican revolt in Cuba aroused ferocious passions in both the United States and Spain. The Hearst newspapers provoked American public opinion and pushed the country toward war. The publisher William Randolph Hearst was certainly more motivated by lingering anti-Spanish sentiment, an animosity no doubt nurtured by his father, than by any feeling of solidarity with the Cuban rebels. William's father, George, had been an adventurer and entrepreneur in California during the gold rush, with little admiration for the Spanish, whom he saw as an obstruction to Manifest Destiny. But there was more to it than nationalism. American investors had a stake in Cuban plantations, and while there was some public sentiment for the Cuban cause against Spain, there were also economic interests at risk.

Because of his own military sojourn in Cuba, Cajal understood the parochial view well and sided with the rebels. The sinking of the *Maine*, he believed, was only a pretext for war and annexation. However, he did come to understand and even admire the Americans a few years later when he lectured at Clark University and traveled in the United States.

Cajal had been surprised by an 1899 invitation to speak in Worcester on the tenth anniversary of Clark's founding. The letter inviting him even contained a six-hundred-dollar check for expenses. The Spanish-American War had just ended, Cajal had recently developed heart palpitations, and

he didn't really want to make the long trip by train to Le
Havre, let alone across the Atlantic into what he feared would
be a hostile environment. But friends and government officials (including the Minister of Education, Marquis de Pidal)
urged him to accept, as much to help restore Spain's honor
after her humiliating defeat as for himself.

Fueled by patriotism, Cajal finally agreed. He and Dõna
Silvería arrived in New York City during the June heat. Almost
immediately after they checked into their hotel the venerable
building caught fire. As it quickly filled with acrid smoke, the
Spanish guests climbed out onto the balcony of their room.
The fire escape leading to the street was so high and rickety,
however, that Señora Cajal couldn't be convinced to use it,
causing her husband to complain, "Who could make a lady
who is timid and nervous, like a good Spaniard, descend
those aerial steps? Luckily the fire engines soon arrived and
quickly extinguished the fire."

Even that startling first event didn't diminish the joy of the
trip. Cajal loved the United States and its people. He was
especially delighted by the sympathy Americans showed
toward Cuban refugees, some of whom were working in the
New York stores he visited.

The Cajals took the train to Boston, and on to Worcester,
in the suffocating heat. Temperature, humidity, and the
inconvenient hour of their arrival left them both with sleep-
lessness and headaches. The next day happened to be the
Fourth of July, an occasion Cajal remembers as marked by
"madmen discharging rifles into the air." He thought the celebration of American independence compared badly even to

the Spanish spectacle of bullfighting and complained about the folly of expressing jubilation by making noise. Receptions, banquets, and tours of the university eventually replaced the gunfire, and Cajal's three lectures on the structure of the cerebral cortex were rewarded with loud applause. The closing convocation was held on July 10, at which time Cajal and four other foreign professors received honorary degrees.

Before starting back to Spain, a romantic detour took them to eastern New York State so that Cajal could photograph Niagara Falls with his Kodak. It is a testimony to his artistic sense of forbearance that, while he found it possible to photograph what he called "the wonderful cataracts," he realized that it would have been "unpardonable now to pause to describe them." Last, they visited the Agassiz collection at Harvard and the Boston City Library. Told that the card catalog contained several million entries, Cajal tested the system when he requested a first edition of *Don Quixote* and rejoiced when he was handed the volume within a few minutes. The librarian who found the book, carefully chosen to help the professor, had spent two years studying in Madrid. In the foreign periodicals room, the same librarian showed him a collection of Spanish newspapers, insulting to Americans, that the librarian claimed bore some responsibility for having incited the Spanish-American War.

Wars and travel might be inconvenient, but when Cajal returned to Madrid his laboratory work continued at the same pace it always had, and his scientific life was stable, productive, and happy. Early in the winter of 1894, a letter arrived from the secretary of the Royal Society in London that slightly disrupted his routine. The European anatomical community had previously honored him, as organizations in Spain had, but an invitation to deliver The Croonian Lecture was something altogether different. The English Royal Society was still the most celebrated scientific organization in the world (Isaac Newton had once been its president), and Cajal was so overwhelmed by this new invitation he wasn't sure he should accept. His unsteadiness in English no doubt worried him, but Charles Sherrington, the British physiologist who would be his host, wrote to tell him that he could deliver the address in French, and urged him to come.

He went to London. A few days before he lectured on his work defining the neuron and "gap" (Sherrington had not yet proposed the term *synapse*), Cambridge University, with proper British solemnity, presented him with an honorary degree. He visited the modern neurophysiology labs of Sherrington and David Ferrier, and spent time with Victor Horsley, one of the first doctors in the world to specialize in surgery on the brain.

It isn't clear whether or not Cajal saw Horsley actually operate on a human patient, but at about the same time the first American neurosurgeon, Harvey Cushing, did and was terrified by the experience. Cushing had been surprised to arrive at Horsley's home and discover him boiling instruments on his own stove. The English surgeon was already widely known for a talk he had delivered in Berlin at the 1890 International Medical Congress. His paper was a report on forty-four craniotomies, operations Horsley had performed (then usually fatal) with only a 25 percent mortality! Cushing's host wrapped his instruments in a tea towel and summoned a cab. The pair then sped to a West End mansion and dashed up the stairs. Horsely "had his patient under ether in five minutes, and was operating fifteen minutes after he entered the house; made a great hole in the woman's skull, pushed up the temporal lobe—blood everywhere, gauze packed into the middle fossa, the ganglion cut, the wound closed, and he was out of the house in less than an hour after he had entered it." This picture of an operation for tic douloureux (severe face pain) is one of the most frightening in the medical literature, and the experience so shocked Cushing that he immediately abandoned Horsley and returned to Johns Hopkins. Cajal would have shrunk from such a scene, but he most likely visited Horsley's small lab at University College where the surgeon conducted animal experiments.

On the other hand, Sherrington proved a good host. He liked Cajal immediately, and admired the neuron theory, but both he and his guest felt that the space between these cellular structures required a name. Cajal had no suggestions. Sherrington was writing material for a new edition of Sir Michael Foster's *Textbook of Physiology* and grew tired of having to invent ways to describe the conduction of an impulse down an axon across a gap to dendrites. The classically trained Englishman first proposed the unattractive term *syndesm*. Fortunately a colleague with a more musical ear suggested *synapse* because it "yields a better adjectival form." This neologism for the gap was first popularized in Foster's new edition, which appeared in 1897.

As much as Sherrington enjoyed Cajal, the entire English household wondered at their guest's habit of locking the door to his bedroom and taking the key with him whenever he left the house. Unsure about Spanish customs, the maid nonetheless had to clean, and so eventually she found another key and let herself in the room. Every shelf and sill was lined with bottles of reagents, slides were scattered about, and a microscope had been set up on the bureau. She ran to fetch Mrs. Sherrington for advice. Puzzled about some problem he had brought with him, Cajal had refused to waste time. This laboratory in his guest bedroom must have amused Sherrington, but it hardly would have shocked him. He had already learned to appreciate the "intense anthropomorphism of Cajal's descriptions" and delighted in the way "he treated the microscopic scene as though it were alive and were inhabited by beings which felt and did and hoped and tried even as we do."

Charmed though his English host may have been by the eccentric Spanish guest, the visit was more daunting for Mrs. Sherrington. At dinner a few nights after the bedroom laboratory was discovered, Cajal punctuated his reasoning during an intense discussion by tearing off small pieces of a soft English dinner roll. In the process, he made a large mound of the dough balls in front of him while building his proof, pulling off one piece of dinner roll per example. Reaching the denouement of his argument, Cajal dramatically swept the pile of his "bullet points" of dough onto the floor, much to Mrs. Sherrington's annoyance. It was this dramatic sensibility of Cajal's that sustained his passionate searching.

Over the next few years, new stains and energetic work enabled Cajal and his contemporaries to map much of the central nervous system in man and other mammals. Unobstructed by the network concept, by the end of the nineteenth century they had successfully described most of the significant mammalian neuroanatomy visible through the light microscope.

In the summer of 1903 during a vacation in Italy, a new idea, one that must have been lingering subconsciously for years, suddenly germinated in Cajal's imagination. He couldn't explain himself why it took him nearly two decades to realize that the photographic process for reducing silver had direct application to the silver-staining technique that had so liberated histology. Several investigators had become interested in the internal structure of neurons, specifically neurofilaments within axons. These filaments proved flirtatious, and the silver salts routinely used in the Golgi method interfered with staining them. Maybe it was the bright sunlight as he lounged on the Italian Riviera that made Cajal think paradoxically of his darkroom and pans of chemicals. On returning to Madrid, he soon perfected a method in which he heated tissue in a solution of silver nitrate, then treated it in darkness with a reducing agent. This technique beautifully stained not only the neurofilaments but also axonal and dendritic endings, because it overcame the reluctance of the Golgi method to penetrate myelinated fibers. Over the next twenty years, he used the reduced-silver technique to investigate neurohistology to the limits allowed by light microscopy and to describe a coherent theory of the nervous system.

One aspect of the reduced-silver discovery must have pleased Golgi: Cajal was able to observe that the ethereal neurofilaments do form a network *inside* neurons.

EARLY ONE OCTOBER MORNING in 1906, a telegram written in German arrived at Cajal's home sent from Stockholm: "*Carolinische Institut verliehen Sie Nobelpreis.*"

In Pavia, Golgi received the same message.

These two men were the first histologists ever to win the Nobel Prize. Behind the scenes, lobbying between their partisans over which one should receive it had been going on for years. Golgi's continued insistence on networks hurt his chances, but the value of his black reaction had always kept his name alive. Still, most of the support went to his rival. In fact it is clear that Cajal was aware he was being considered for the prize. In October of 1904, he wrote to thank Professor Leon Corral y Maestro, who had most recently nominated him, admitting that "this nomination has previously been made by other Spanish and foreign universities with little success." Indeed, both Golgi and Cajal had been proposed every year beginning in 1901. On the morning the cables arrived in Madrid and Pavia five years later, neither of them was particularly troubled that they would share the Nobel Prize, despite their alienation.

Cajal's first thought was typically modest: "How could I justify the preferences of the Carolinian Institute in the eyes of so many outstanding investigators who had been passed over." About the same time, Golgi wrote to Count Karl Morner, the rector of the Karolinska Institute, acknowledging a "satisfaction at finding myself accorded this honor together with such a scholar as Mr. Ramón y Cajal," but he didn't really mean it, any more than he looked forward to going to Stockholm.

Neither of the laureates was in good health. Golgi frequently had minor ailments during these years, though he admitted that on this occasion he was nervous and wanted to be left alone more than he usually did. Cajal had continuing heart problems, made worse by his anxiety over the occasion and the effort of another long trip. In fact, even knowing that the Nobel rules require the laureates to attend personally, he had initially answered Morner's telegram by saying that he was too sick to make the journey. But of course he did, and he was there ahead of Golgi.

The Italian's arrival at the Stockholm train station is a scene that condenses the intertwined lives of these two men. Late on December 8, 1906, Santiago Ramón y Cajal stood amiably on the platform, well wrapped in a heavy Spanish cloak purchased especially for the trip north, to welcome Camillo Golgi at the Stockholm railroad station. For the proud Spaniard, this was an obvious olive branch. In his *Recollections*, Cajal never mentions the moment, but another Italian scientist who also met the train remembered:

> There was no better occasion to put a cap on the past and acknowledge the merits of both of them. . . . And it would

have been truly agreeable if a gesture of reconciliation had originated from the older of the two, because Cajal, with his presence, had in a certain way encouraged him to do so. But the ice was not broken either that evening or in the subsequent unavoidable encounters during their stay in Scandinavia.

In any case, after briefly greeting those who met him, Golgi simply walked off the platform and went directly to his hotel. His wife, Lina, had slipped and fallen on the frozen ground getting off the ferry in Trelleborg, injuring herself so badly she had difficulty getting to her feet. Because her husband wasn't feeling well either, their trip was delayed for a few days along the route. These physical problems added to his existing anxieties may help to explain why he ignored Cajal's gesture. For the Spaniard, this would have been an insult to him personally and to his country. When he discussed his own feelings about winning the Nobel Prize, though, Cajal observed much more than the moment:

> Every well-bred person must be grateful for them [such honors] and remember them. But we Latin people are extremists in everything. In contrast to the moderation and coolness of the northern peoples, we lack the sense of proportion and of balance. . . . Like vehement and rude friendship, among us fame bruises while it caresses; it kisses but it crushes. It deprives us of the ease of custom; it disturbs the peace of the spirit; it restricts the sacrosanct freedom of the will, turning us into the target of impertinent curiosity; it endangers humility, compelling us continually to think and speak of ourselves; and, finally, it

alters the course of our lives, twisting it into capricious and useless meanderings.

This lyrical opinion about regional differences of character was a core belief and made its way into Cajal's book *Vacation Stories*, short stories he called science fiction and published under the name Dr. Bacteria. The collection would have gone unnoticed save for the fame of the writer, whose purpose was to instruct his Spanish readers about how science worked. In addition, the stories reveal their author's then-suspect liberal views on politics, religion, and class, and while they fail as art they are useful for this reason alone. In "Secret Offense, Secret Revenge," a wholly unbelievable account of indiscretion and retaliation, Cajal draws a great distinction between passionate Latins and "prudish daughters of the north." Although Golgi was Italian, he was northern Italian, and he was nothing if not personally cool.

The banquets and receptions surrounding the Nobel Lectures seem designed to unnerve the laureates, but Golgi needed no help. He was a wreck. Absent from full-time research at the laboratory bench since 1893 because of his duties as rector and uncertain of the most recent science, he was agitated further by having to rewrite some of his talk, which was to be delivered in French.

Golgi spoke first, and the Nobel Lecture was his final undoing. Although an occasional gasp could still be heard from defenders of the reticular theory, neurons were by then almost universally accepted. Golgi really didn't have to get into this matter at all; he could simply have described the black reaction and his own anatomical contributions, passing over the entire controversy. In a bland introduction that detailed the complexity of the nervous system and praised the contributions of both men, the rector of the Karolinska Institute had given him every opportunity to do just that.

But he did not. His lecture was argumentative and tautological. In the first paragraph he reiterated that he had "always been opposed to the neuron theory" and then obliviously claimed that "this doctrine is generally recognized to be going out of favor." In one breath, he had insulted nearly every scientist in the hall. After restating his defense of the network, he absurdly concluded that the neuron theory was wrong because it was incompatible with his proof. He denied the existence of cerebral localization and maintained that "the idea of the protoplasmic processes having a nutritive function is, in my opinion, corroborated by the well-known facts." By the middle of the speech he was, according to recollections of some observers and evidence from the manuscript itself, nearly raving.

He attacked the solid anatomical evidence that neuroblasts develop into neurons; he ridiculed the concept that mature neurons are a single cell composed of body, axon, and dendrites; and with a special virulence aimed at Spain, he attempted to overturn the law of dynamic polarization. While calling Cajal's insight brilliant, he promptly dismissed the idea that neurons conduct in one direction, from dendrites to synapse, saying this was only one interpretation. He ignored the enormous contributions made by other scientists: Kölliker, Waldeyer, Forel, Retzius, and His. Retzius sat in the audience with Solomon Henschen (who had described the fine structure of the visual cortex), and both "looked at the speaker with stupefaction." Golgi ended by embracing Flourens's antiquated notion that there exists a unity of action in the nervous system, adding he didn't care if the theory was an ancient one.

Golgi never understood what he had done, and no one in the audience ever forgot it.

Born during an era when dozens of the greatest observers in natural science were living, Camillo Golgi remains a compelling figure in neuroscience. And his curiosity traveled far beyond the nervous system. He discovered the sensory ending in tendons that bears his name, and he first saw the now vitally important subcellular endoplasmic reticulum and Golgi apparatus, structures key to understanding cellular metabolism. He studied malaria, helping to determine the life cycle of the malarial parasite. Golgi was a dedicated teacher whose students were devoted to him. Toward the end of his career, he was respected not only as a scientist but also as an administrator and politician.

Yet there is something troubling about Golgi. It is to be found both in the fragments of his own writing and in his biographies: characteristics perhaps rooted in him when he was removed from his father, struggled socially as a young man, or was judged not qualified for a professorship at Pavia. Though Golgi knew Freud (who early in his career did some excellent anatomical research), he was not a man to have seen his life in Freudian terms, and he left no autobiography by which to explain himself. The picture that emerges from what has been written about Golgi the man reveals a withdrawn, narcissistic, and melancholy person. In too many places the papers and letters that he did leave are punctuated by references to *my* black reaction, *my* method, *my* discoveries, *my* theory. Even the accounts written by his admirers show Golgi to have required continual reassurance in order to guard his position in the intricate Italian scientific hierarchy. His most intimate relationships were with his wife, who adored him, a niece, and his own intellect.

In contrast, Cajal's lecture was about science, not ego. He simply described his work. With humility, he acknowledged the research of others, particularly His, whose original studies of neurogenesis he stoutly defended. The session ended with the long applause that expresses acceptance and admiration.

In a truly contradictory tribute the day after Cajal and Golgi spoke, the warrior Teddy Roosevelt received the 1906 Nobel Peace Prize. The regular military officers of the U.S. Cavalry, who did most of the real fighting, found Roosevelt's gallop up a hill in San Juan leading the Rough Riders both reckless and unnecessary. Privately they called it the "schoolboy charge." The escapade was political, not military, undertaken to satisfy Roosevelt's rambunctious ambition. The job he eventually sought in the McKinley administration for his political work as a loyal Republican was delayed, Henry Cabot Lodge wrote, because "there is a fear that you will want to fight somebody at once." Even as assistant secretary of the Navy, he had been looking for trouble from the very moment of his appointment. In 1897 he wrote to a friend, "In strict confidence . . . I should welcome almost any war, for I think this country needs one." After being at his new post for only two months, Roosevelt gave a speech at the Naval War College calling for expansion of the fleet and preparation for a "*necessary* war." In spite of his personally combative attitude and unflagging belief in the Manifest Destiny of the United States, Roosevelt did later help negotiate an end to the war between Japan and Russia in 1905. For this more than anything else he received the Peace Prize.

Though he tried, Cajal was unable to defend Roosevelt's aggressive personality, or his adventures in Cuba during the Spanish-American War that won for him so much admiration in the United States. In his *Recollections*, Cajal later wondered rhetorically, "Is it not the acme of irony and humor to convert into a champion of pacifism the man of the most impetuously pugnacious temperament and the most determined imperialist that the United States have ever produced?"

The new century understands no better how to tame this temperament—what the moralist Crane Brinton has called "the will to shine, or the will to howl, if not the will to power"—than did those who lived in 1900. Deciphering form and function of the mysterious individual cells that make up human brains, Cajal realized, doesn't explain their final expression in the moral (or immoral) behavior of individual people. "Accursed House," another didactic tale from his collection *Vacation Stories*, ends with Cajal's complaint that "*homo sapiens* only philosophizes in his spare time. He is still

too low on the intellectual scale, too utterly dominated by the
reflexes of his stomach. Thought in his brain is a bird of passage, an irritating guest who interrupts the endless traffic of
interest and greed."

TEN

Higher Magnification

CAJAL NO MORE SAW the actual synaptic space than his great antagonist actually saw the network. Thanks to the tools he inherited from Golgi and others, a compulsive capacity for hard work inherited from his father, and the wonderful three-dimensional imagination passed to him perhaps by his mother (Cajal is quiet on the subject of his mother; in that respect, he himself was the chauvinistic Spanish male of his era), he deduced the synapse. Wherever these abilities came from, they flourished in him. Once he had made his crucial observations about the cerebellum and retina and understood the fundamental anatomy of the neuron, he intuited that the hand of the presynaptic axonal ending exactly fit the glove of the postsynaptic dendrite.

While a general acceptance of Cajal's explanation for the synapse didn't suddenly make people behave better, it soon opened the way for more sophisticated physiological research. Émil du Bois-Reymond was a German of French ancestry, a contemporary and friend of Helmholtz, and eventually professor of physiology at the University of Berlin. An ingenious researcher, he manufactured most of his own instruments and had already brought physics to the study of the nervous system. By the middle of the nineteenth century, du Bois-Reymond had characterized the action potential, suggesting that it is not current flow itself that transmits information but electrical depolarization across neuronal membranes, followed by a return of the cell to its stable resting state. That is, the cell turns on and off.

Just as the century turned, his student, Julius Bernstein, another Swiss-German, added biochemistry to the system. Bernstein proposed a theory that selective permeability of the neuronal membrane to potassium excited the action potential. Soon physiologists all over Europe were investigating movement of sodium and potassium into and out of cells, and beginning to understand the chemistry that causes muscles to contract and neurons to depolarize. At Cambridge in 1905, John Langley suggested the idea that axons contained a chemical "synaptic substance" that might stimulate a receptor on neighboring dendrites. We now know that he was right. There are many such substances, now called neurotransmitters, liberated by an action potential into the gap Cajal found, whose purpose is to instruct the next cell.

Like all tools, the achromatic light microscope has limits, and by early in the twentieth century the limits of optical resolution had been reached. It was left to the next generation of electron microscopists, biophysicists, and biochemists to actually see the membranes and the synaptic space itself, to understand membrane depolarization, and to find the neurotransmitters.

Just as better technology permitted Cajal and his contemporaries to recognize that what earlier observers had called granules were really cells, and that their fibers were axons, so did electron microscopists see more deeply into the structure of the neuron. In 1926, Hans Busch, a German theorist, first suggested that electrons might be focused in a magnetic field to make a lens system. Technology galloped along quickly between the World Wars, and by the 1940s, practical electron microscopes were for sale. These instruments work in a way opposite to light microscopes in that, rather than the viewed object's absorbing light, the atomic nuclei in the specimen *scatter* electrons, thereby producing the image. But most organic material scatters electrons so weakly that the objects being observed could not be distinguished at first from the supporting film that held them together and allowed them to be cut. When early electron microscopists looked at biological specimens, it was as if they were trying to see a piece of clear tape stuck onto a window.

Staining with heavy atoms (osmium, tungsten, or uranium, for example) increases this scattering. Better contrast outlines the components of tissue, so that the specimen stands out from the embedding material. These stains, developed especially for electron microscopy after World War II, plus the ensuing refinements for fixing and cutting biological tissue that followed, soon began to produce pictures of the subcellular structure of cells. At Rockefeller University, Sanford Palay was the first electron microscopist to combine these technical advances with enough understanding of the nervous system to see the fine structure of the synapse itself, and to confirm Cajal's deductions.

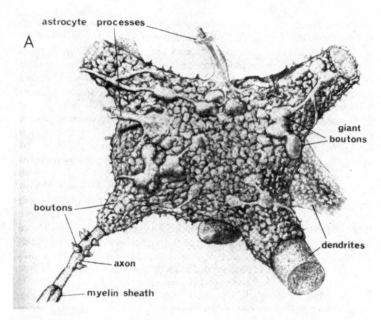

Synapses completely covering the surface of a motor neuron cell body. *From Poritsky, "Two and Three Dimensional Ultrastructure of Boutons and Glial Cells on the Motoneuronal Surface in the Cat Spinal Cord."*

Light microscopes showed axonal endings conforming to dendrites, but anatomists were now able to find the most intimate details of these meetings. Photographs obtained through the electron microscope reveal that the surfaces of neuronal cell bodies and their dendrites are plastered with many more bouton endings of other axons than silver stains disclose. Most neurons have about a thousand contacts, and each of the very popular Purkinje cells may attract as many as *eighty thousand.* Thus an enormous amount of information goes into a neuron simply to turn it on—to generate the action potential that ignites its own synapse. The myelinated part of these endings halts just short of its bouton, and the bare axon then ends as a presynaptic membrane, like the stripped end of an insulated copper wire. This terminus is separated from the other side, the postsynaptic side, by a synaptic cleft— that *gap* that started all the argument in the first place. Many electron microscopists saw this much of synaptic anatomy not long after Palay's first reports appeared in 1953. But the fine structure is much more ornate and much more interesting in ways that could only be guessed at in the nineteenth century.

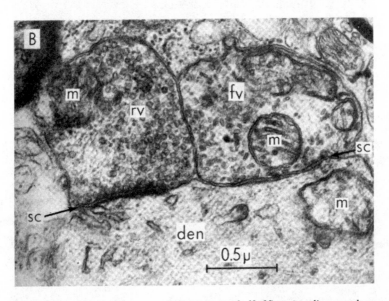

Electron micrograph of two synaptic boutons (top half of figure) ending on a dendrite (bottom half of figure, labeled "den"). The synapse on the left is excitatory: The vesicles are spherical (rv) and the staining on each side of the synaptic cleft (sc) is dark. The one of the right is inhibitory: The vesicles are ellipsoid (fv), and the staining of the cleft is much lighter. *From Grey, "Electron Microscopy of Excitatory and Inhibitory Synapses: A Brief Review."*

Those earlier neuroanatomists and physiologists knew that an electrical impulse shot down the axon and somehow stimulated the next cell, and Cajal had even hinted that synaptic transmission might be chemical. But the actual encounter was beyond their technology. Palay and other neuroanatomists learned that the typical axon terminal is actually a chemical factory, producing a variety of compounds and storing them in packets. These synaptic vesicles come in several sizes and shapes, and the form of the packet predicts its function. Further, the architecture of the membranes on either side of the cleft adds more information about whether that particular synapse is excitatory (making the cell more likely to fire) or inhibitory (encouraging it not to fire).

By the 1960s, this varied fine structure of synapses was already well understood, but the biochemistry of transmission was not. Early in the twentieth century, chemical transmission across the synapse, especially at the motor end-plate where the bare axon first branches and then stimulates muscle fibers, was only a theory. In a famous 1921 middle-of-the-night lab exercise remembered as the two-toad experiment, Otto Loewi electrically excited the cut vagus nerve to a rapidly beating isolated frog heart, causing the heart's rhythm to slow. More important, the perfusion fluid he collected from the first frog caused slowing when introduced into a second isolated heart—clearly a chemical event. Using powerful twentieth-century electronics, an enhanced understanding of neurochemistry, and much more sophisticated microelectrode recording from single cells, scientists had established the basic nature of synaptic transmission by the time Sputnik was launched. But some of the chemical details were still sketchy.

The research I was participating in at the University of Chicago during these years was part of a much larger effort to understand the specific biochemical nature of the synapse. Laboratories all over the world were involved in this campaign, but one of the major forces behind work on the chemical synapse in the United States was Julius Axelrod at the National Institutes of Health. A short, good-natured, animated man, he was blind in one eye from a lab accident by the time I met him—a deformity not concealed by an opaque lens in his glasses. He had Cajal's energy, focus, and passion for finding the secrets hidden in the nervous system.

Exactly what were these little packets of chemicals, and how did they boat across the tiny pond of the synaptic space? Trying to find out, we investigated the pineal gland of vertebrates—the same minute pineal body to which Descartes had assigned metaphysical authority. The pineal is a curious structure both anatomically and chemically, and happened to provide a convenient neural system that could easily be manipulated. Axelrod was coordinating efforts that exploited opportunities offered by peculiarities in systems such as the pineal to expose the details of synaptic transmission. The next year, Axelrod shared the Nobel Prize for his work on the biochemistry of neurotransmitters.

Ancestral vertebrates had a third eye, in the middle of the forehead, for looking up—functional in finding the surface if you're swimming deep in the ocean but inconvenient otherwise. This eye persisted in bony fishes, older amphibians, and early reptiles. "But by the Triassic times," the comparative anatomist Alfred Romer observed, "median eyes had apparently gone out of fashion." Small wonder. Although the dorsal forehead eye vanished in higher vertebrates, the pineal gland is its remnant and remains hooked up to the visual system. In addition, the pineal manufactures melatonin, a hormone related to the biological clock and (at least in rodents) to sexual function. Because it indirectly still perceives light, pineal production of melatonin can be influenced by light or dark, and the visual effects on hormones in rats can be measured. The biochemistry of melatonin was also known, so the impact of light on the gland itself could be gauged. As a consequence of these characteristics, several leading neurobiologists had found the pineal a useful tool for investigating synaptic transmission.

The minute details of the several biochemical mechanisms for synaptic transmission are beyond us here. Evolution often produces redundancy and builds several ways for doing similar things. Claude Shannon, the father of information theory, puts it this way: "That the brain has ten billion neurons probably means that it was cheaper for biology to make more components than to work out sophisticated circuits." Complexity and multiple systems needn't be a barrier to understanding the fundamentals, however, and we can sketch a basic description of the chemical synapse well enough to complete Cajal's picture. In addition to dopamine (the lack of which produces Parkinson's disease), other proven transmitters, substances that cross the synaptic cleft between the axon and dendrites, are norepinephrine, serotonin, acetylcholine, and some amino acids. More recently, several other classes of compounds have been identified as sometime transmitters, including endorphins, adenosine, and even a few familiar gases such as nitrous oxide and carbon monoxide.

Biochemists mapped the concentrations of the first proven neurotransmitters in specific brain cells and tracts using histochemical techniques that cause fluorescence. Once located, high concentrations of these now eerie greenish-stained molecules and their metabolites might suggest what they were doing there. Eventually, as in the case of dopamine, serotonin, and acetylcholine, for example, researchers were able to link these chemicals to specific diseases in patients: Parkinsonism, depression, myasthenia gravis.

Because many different kinds of synapses and transmitters have now been found, and since there is no common biochemistry completely describing them all, generalization about a typical model is difficult and not completely accurate for all the cases. Yet, an example can give an idea of how many of them work.

Classical transmitters are manufactured and then stored in the vesicles Palay and his co-workers originally found on the presynaptic (axonal) side of nerve endings. When an action potential jumps down the axon, changes occur in calcium concentrations at the presynaptic membrane, resulting in a release of transmitter into the synaptic cleft. A variety of mechanisms exists to bind these transmitters to receptor sites on the postsynaptic (dendritic or cell-body) membrane. The stimulation that follows may be either excitatory or inhibitory; that is, the electrical gradient across the postsynaptic membrane may increase or decrease. It is the sum total of all such inputs—plus their exact location on the receiving cell—that determines whether or not that cell depolarizes, producing its own action potential and firing down its axon, in turn stimulating the next neuron. Once the receiving cell has been aroused, the chemical transmitter is taken up again or metabolized back into its components, and the stimulating neuron returns to its normal resting potential; it turns off. In summary, the steps are: neuronal depolarization, action potential, presynaptic liberation of transmitter, stimulation of the postsynaptic membrane (one of at least a thousand on every neuron), reuptake or metabolic breakdown of transmitter, and return to stable resting potential. What happens to the stimulated neuron depends on the total influence of all the events on its dendrites and cell body. While the biochemical details of different kinds of synapses vary depending on the type of transmitter involved, they all fundamentally rely on an electrical gradient across a membrane.

The basic structure of these cell membranes appeared early in evolution, and is therefore similar in the neurons of all vertebrates and invertebrates. Cajal's English host Charles Sherrington first investigated much of this neurophysiology, and his student John Eccles deepened the exploration when he mastered the technique of recording inside a single cell. To do this sort of biophysics, Eccles inserted a fine glass microelectrode only half a micron (.0005 millimeters) across at the tip and measured the changes in electrical potentials *inside* the neuron. For all their original work, Sherrington won the Nobel Prize in 1932 and Eccles in 1963. Neuronal membranes maintain their electrical gradient (just as dissimilar metals make a battery), using energy derived from adenosine triphosphate (ATP) to transport ions in and out of the cell. Depolarization of the cell membrane generates an action potential that usually liberates one of the several different chemical transmitters.

There are contrarians, however. At some synapses, called gap junctions, the pre- and postsynaptic membranes are so close together that stimulation is directly electrical, skipping the chemicals altogether. Just like a spark plug, electricity jumps across the gap. This important exception to the more classical model has implications. Regardless of the specifics, the coding of all information in the central nervous system depends on maintaining a stable electrical gradient across membranes until an action potential occurs, and the next neuron is thereby stimulated. But that next neuron gets lots of bossing around from its thousand or more synaptic inputs, some encouraging it to fire an action potential (stimulation) and others insisting that it not fire (inhibition). The total number and type of inputs determine what the cell does. This is true for relatively simple sensory messages—touch, for example—and motor responses, as well as the much more complex functions of vision, language, and memory.

While Cajal could not have exposed all these intricacies

with the light microscope, he was correct in his basic descriptions of neuronal anatomy, dynamic polarization, and synaptic transmission. Perhaps the most remarkable part of his work was not the anatomical descriptions themselves, which even now are elegant and beautiful, but that he could deduce so much from the outlines that he did see. In this way, it might have troubled him to admit, he reasoned in the manner of Aristotle and the other ancients he dismissed. He deduced function from the limits of what he could see. The moon looks a lot different from earth than it did to Neil Armstrong when he was standing on it. Cajal, knowing from experience that theories are bound by observational methods, would probably not be surprised to discover the library of new information that was concealed by his simplification of the synapse into the word *contact*. It is perhaps unexpected, though, that some recent, more detailed methods for understanding the synapse lead to the conclusion that Golgi, at least in some theoretical ways, was right, too!

Golgi's main scientific error, we have seen, was in a devotion to his preconceived idea about how the nervous system *ought* to work. He clung to the network theory because he believed it gave an evolutionary advantage to the mammal with the most complex nervous system—us. This is the holistic and metaphysical idea of Flourens served on a bed of Darwinism. The superiority of a nervous system designed in this reticular way, constantly informing the whole organism about its environment using unlimited random-access memory, seemed obvious to Golgi. Had he known about American freeways, he might have changed his mind.

The movement, storage, and retrieval of information require turning cellular units on and off. An entirely connected nervous system would sometimes behave like the constant flow of traffic on a freeway at rush hour: too many cars, too many bottlenecks, too many accidents, too few exits. In the 1940s, Sherrington used this same analogy in discussing Cajal's law of dynamic polarization. "He showed, for instance, that each nerve path is always a line of one-way traffic only. . . . The nerve circuits are valved, he [Cajal] said, and he was able to point out where the valves lie—namely where one nerve cell meets the next one." Without valves (or stoplights) eventually movement would become hopelessly snarled and stop. Synapses add stoplights at the on-ramps (in this analogy, gating information) to direct traffic flow. There are even modern theories that add high-occupancy vehicle (HOV) lanes to the traffic flow.

According to some current speculation about what are now being called neuronal networks, certain kinds of information (ideas or events well learned, highly emotional, or very important) might depend on groups of cells with special relationships. How the term *neuronal network* would please Golgi! Clusters of fibers around certain cortical neurons discovered in the 1930s led to a theory of intercommunicating vertical columns in gray matter. Thirty years later, such a columnar arrangement in the sensory cortex of a rat, termed a "barrel," proved to contain all the sensory neurons needed to recognize the stimulation of a single rodent whisker. So populations of cells with similar form and organization do function together as a sort of network. While these cells are individuals, and not directly connected as Golgi would have had it, the synaptic communication within the networks is heavily influenced by their history of activation. Having once been turned on, the cell is more likely to be activated in the future. This phenomenon is termed plasticity and is certainly related to learning. The essential element in modern network theory is not the function of a single neuron but the synaptic influences of all the cells within the network on one another.

Similarities between biological neural networks and computers have resulted in computational abstractions known as artificial neural networks, mathematical schemes that enable computer systems to perform tasks outside of traditional statistical rules. Such abstractions are now popular in building theories of artificial intelligence and in modeling the neuron itself, as well as in studying how systems work within the nervous system.

The final concession current neurobiology must make to Golgi is the gap junction. According to Michael Bennett, an expert in electrical synapses who teaches at Albert Einstein College of Medicine, these structures "constitute a small but respectable minority of synapses in the mammalian brain, as well as the brains of lower forms." As we have seen, the pre- and postsynaptic membranes are so close at gap junctions that the chemical nature of synaptic transmission surrenders to direct electrical stimulation of one cell by another. Golgi would have rejoiced to know that minute protoplasmic viaducts called "connexins" actually bridge the pre- and postsynaptic membranes at gap junctions. Even more wonderful, the connexin proteins are manufactured in the endoplasmic reticulum, transported through the Golgi apparatus, and inserted into the cell membrane to form these sites. Although the advantage of the arrangement is uncertain (Bennett has said of electrical synapses, "They can do many things that chemical synapses can do, and do them just as slowly"), it does raise the question of how close is touching. Golgi believed that he had reasoned his way to the answer, and Cajal thought he had seen it. Both of them may have been correct. At the end of the nineteenth century, when understanding of the synapse was limited by the resolution of silver stains and light microscopes, truth was, in part, a matter of intuition and manner. At the same moment, the pragmatic Oliver Wendell Holmes reasoned that the "law is nothing more or less than what judges do." Holmes did for the law what Darwin did for evolution and Maxwell did for gases, writes Louis Menand, when, "he applied to his own special field the great nineteenth-century discovery that the indeterminacy of individual behavior can be regularized by considering people statistically at the level of the mass." We hope for more certainty from the laws of science, but we are just as often disappointed.

ELEVEN

Recollections of Two Lives

For both of them in the end, who was right depended, at least in part, on a matter of intuition and manner. This was true in 1889 and remains true today. Despite their competition, Cajal and Golgi shared many more similarities than differences: the positivism of their day, the scientific method, patriotism, devotion to family and students, dedication to the nervous system, a respectful mistrust of religion. Finally, of course, they shared the Nobel Prize. But whereas Cajal saw the world through the eyes of a passionate, liberal south Mediterranean, Golgi saw the same world with the colder eye of a conservative northerner. Looking down similar microscopes at the same tissues, they each saw a different nervous system. Regardless, the work of one depended on the other. Cajal summarized the entire conflict when he wrote, "What a cruel irony of fate to pair, like Siamese twins united by the shoulders, scientific adversaries of such contrasting character!"

After leaving Stockholm in 1906, these two master observers of the nervous system never saw each other again.

The Nobel Prize finally won Golgi the status for which he

had yearned, as well as a substantial purse of research money (his half was about twenty-three thousand dollars), but it did not win him peace even in Italy. No matter how noble the effort, he simply attracted controversy, and his political activities as rector of the University of Pavia and in the Senate made him a target. His efforts to renovate the university were obstructed by partisans of an expanded State University of Milan; there were misunderstandings about his use of malaria research funds from an American grant; the great political upheaval in Europe polarized his countrymen; and even when neurons were universally accepted, he adamantly defended the network.

The threat from Milan was championed by an aggressive surgeon, who, according to Golgi's student Antonio Pensa, was a "strong-willed, smooth-talking, eloquent but somewhat pompous, academic" named Luigi Mangiagalli. When a wealthy engineer died in Milan leaving a bequest for the city to build a new medical school, doctors in Milan naturally found it in their best interest to proceed with it. A second institution only a few kilometers away would certainly have harmed the University of Pavia, and successfully warding off this menace became a primary occupation for Golgi in his last years as rector. It also created another enemy for him in the person of the single-minded Mangiagalli and added more political intrigues to his crowded roster of conflicts.

He filed his final spirited defense of the reticulum following the publication of Georges Marinesco's 1909 book *La Cellele Nerveuse.* Fully embracing the neuron theory and attacking the network, this text, in some of its claims, according to Paolo Mazzarello, "was tantamount to accusing Golgi of being delusional." Marinesco had once passed through Pavia and had briefly visited the Laboratory of General Pathology. He wrote that while he admired Golgi's "beautiful preparations," he never saw networks or any connections between dendrites and blood vessels. In his reply, Golgi didn't modify his defense of the network at all, but he did admit that the role of dendrites might include not only trophic functions (requiring the unseen relationship to blood vessels) but also a "specific function that is thought to belong to the cell body." This specific function, of course, could only really have been synaptic input. He didn't even remember the Frenchman's visit to his lab, although he didn't deny it had happened, arguing that just because Marinesco had missed seeing the reticulum, "the fact is not less true for this and I can assure that other scientists have not only confirmed the observation, but have obtained very convincing preparations of their own."

Golgi continued to work in his lab, and to defend what had become indefensible, even as the world headed toward war. Seeking new markets and colonial power in the fall of 1911, Italy invaded Libya, warning Turkey not to resist the expansion. The annexation of a colony in North Africa ignited unusual enthusiasm in northern Italy, and a rejuvenated unquestioning nationalistic pride anticipated the jingoism of World War I. The political and patriotic side of Golgi's nature, at times attracted to rising Italian fascism, enthusiastically supported the "glorious events in which we have had the privilege to assist during this year," and he undemocratically proclaimed that he was "proud to belong to the ruling class."

As international hysteria grew and the inherent weakness of the Triple Entente (Italy, Germany, and Austria) was exposed, Italy divided itself into those interventionalists who favored the war and those neutralists who wanted to stay out of it. These boundaries were sometimes fluid, though, and in one notable conversion Benito Mussolini was expelled from the neutralist Socialist Party, founded the newspaper *Il Popolo d'Italia*, and immediately became a vehement proponent of Italy's joining the conflict as a member of the Entente. Yet, the new Italian interests in North Africa coincided with those of the French, not the Germans or Austrians, and eventually Italy was to send young men into battle on the side of the Allies.

As a medical student, Golgi had been swept up by the nationalism of Italian unification and had approved when the revolutionary patriot Giuseppe Garibaldi encouraged the young men of Lombardy to fight against Austria. He remained consistent and favored intervention, which came soon enough. By 1915, disruption of the public health by war forced the conversion of the University of Pavia Hospital into military wards. The seventy-two-year-old Golgi helped in the renovation and used his political influence to provide funding for the effort.

The next year, his old friend from Bizzozero's lab, Nicolo Manfredi, died at eighty. Golgi's own troublesome health became more delicate. At war's end, the university required him to retire when he reached the mandatory age of seventy-five. In a final blow to his fragile pride, the administration conceded that, while he might sit in on the meetings of the faculty senate, he could no longer vote.

Golgi's life, while glimpsed through his letters and in his scientific writing, emerges primarily through biographies and the memoirs of others. The portrait is necessarily incomplete. Cajal, on the other hand, painted his own likeness in vivid Spanish colors. In 1917, the second part of his *Recollections of My Life* was published. Though this volume mainly explained his anatomical discoveries, Spanish readers snatched it off the shelves and reviewers praised it even in Italy. None other than Golgi's great partisan and biographer Mazzarello, while calling Cajal's writing style "romantic, argumentative, and aggressive" conceded that it was "effective," and moreover that "it was a stinging and brilliant story that in many places appeared to have been written more with the hand of an artist than of a scientist." Most of Mazzarello's claims for both Golgi and Cajal are persuasive, but none of them fully explain what memories now remain of these scientific antagonists. Golgi wrote the only real epitaph he left—an argumentative diatribe—and preached it for the world to hear from a lectern in Stockholm. Cajal wrote poetry and published it in *Recollections of My Life*.

Cajal returned from Stockholm to the same life he had left in Madrid. True, he bore the distinction of the Nobel Prize, and his half of the money supported a better-funded laboratory, but he maintained a distance from administration and politics. Even when the leader of the Liberal Party, his old friend from the cafe Segismundo Moret, tried to capitalize on his reputation by appointing him Minister of Public Instruction, he refused. Wisely leaving this work to others, Cajal went back into his lab to do what he had always done. Still unable to separate himself from his Italian twin, he devoted several years to perfecting a stain for studying nothing other than the intracellular Golgi apparatus (a sort of transfer station for cellular secretions), an organelle Golgi first discovered in neurons but that is found in all cells.

As Cajal began to age and to slow down scientifically, the controversies lost importance for him. Following a minor squabble concerning neurofibrils, which was nonetheless heated, Cajal wrote that "the morbid desire to assert and to make prominent one's own personality, to be original above all things, wreaks ruin in our time." A profound curiosity motivated Cajal, who was always much more intent on fighting with nature "to wrest new truths from it" than he was in fighting "against other men to defend the truth." From his lifetime of experiences with the egos of investigators, he concluded that success in science "dislodges some deeply rooted error and that behind it is usually concealed injured pride, if not enraged interest." But even as he withdrew from scientific battles, his interest in the nervous system flourished and sustained him.

Always attracted to the beautiful, Cajal now rediscovered the exquisite writing and drawing of the French entomologist J. Henri Fabre on the subject of insect behavior. This modest man, whom Darwin himself called "the incomparable observer," spent most of his life watching insects in the fields near his home. Fabre was more at home with these creatures than he was with people, and sometimes addressed them as "dear insects" when writing about them. In part because he noticed that the Languedocian Sphex, a hunting wasp, only attacked the female Ephippiger cricket, never the male, Fabre assigned to insects the faculty of "not only instinct, which is like an innate understanding, but a certain amount of discernment, enabling them to triumph over unforeseen accidents." With the Frenchman's further compelling descriptions of wasps' cleverly designed nests as a guide, Cajal launched a futile search for neural anatomical sites that might represent the focus for instinctive behavior in wasps and ants. He finally gave up, concluding that insects, like higher forms, are guided by all the sensory data they can collect, and that no special cellular anatomy is associated with "internal impulses." This was, of course, before Oswald Avery determined in the late 1930s that DNA is the substance of genetics.

Not long after this, Cajal's eldest son, Santiago, died. This young man, named for his father, and the child who most resembled him both physically and intellectually, had become ill with typhoid fever in his youth. He was never again well, and died at thirty-two. Unable to pursue a profession, a sadness Cajal forbears to discuss ("Why call up pains of which the only mitigant lies in forgetting!"), Santiago had spent his short, unhappy life as a bookseller. In an effort to make his son more self-sufficient, the father tried several strategies, and the final effort returned him to his passion for photography. In a hope that his son might expand his business into editing and publishing, Cajal wrote the initial chapters of a book that was to be their first project together.

George Eastman had marketed roll film by this time, making photography more available to everyone. Cajal's book about color photography, which he edited and published himself after his son died in 1912, focused on the fundamental physical principles of the art. "I owe to photography unutterable satisfaction and comfort. . . . I wish to pay tribute to that which I have so greatly loved, that which is so worthy of the devotion of every spirit that is sensible. . . . about the beauties of nature." Cajal had always found a procedural similarity between histology and making photographs. He admonished photographers to experiment with colors rather than "adhering to receipts and formulae, like a carpenter," and lectured that the remedy for failures was careful attention to the variables involved in the process. These were the same standards he applied in his laboratory, and the book was a simple and successful explanation of the principles of color photography, including strategies for avoiding mistakes.

His son's death coincided with the onset of world conflagration, and these events seem to mark the beginning of Cajal's slow decline. He exposes a real anguish over the loss of his child when he later recalls, "Like every old man, I also feel all those poisoned wounds of the heart and brain. They are the knocks of time at the door."

For the first time in 1914, an existential note can be heard in Cajal's recollections. He continued to work but was physically unwell and unhappy. Did it matter, he wondered, what biology produced during this moment perverted by "bloody crises of civilization, [when] only those sciences are recognized which are placed with shameful submission at the services of the destroyers of peoples?" Wherever he looked he found cannon, airplanes, gases, poison, epidemics. Laboratory instruments and reagents were scarce or unavailable, and a curtain fell on the work that had characterized his long engagement with the synapse. Most of his scientific contemporaries were dead when it rose again, and Europe was in shambles.

CAMILLO GOLGI RETURNED to research after he retired as rector and worked until shortly before he developed an upper-respiratory-tract infection that led to septicemia (infection in the blood). On a difficult trip to Rome in 1924, where Golgi had advocated for a new San Matteo Hospital, his health dramatically worsened. By May of the following year, when the University of Pavia held a celebration of the beginning of its eleventh century, he had improved slightly, but was still unable to go out. King Victor Emmanuel III, who was in Pavia for the event and learned of the old scientist's decline, sent his aide-de-camp with personal greetings. Though confined to his house, Golgi could watch the parade led by the monarch and was able to return his wave as the procession passed below his balcony.

Golgi's wife and niece Carolina were with him constantly as he rapidly failed. One of his former students, Edorado Gemilli, also came to the bedside. This most raucous of medical students had begun his career in Golgi's lab as a capable anatomist and a loud atheist. On this day, as a Minor Franciscan friar, he unsuccessfully attempted a deathbed conversion of his master.

Not even genius and a temperate life spent at the microscope could prevent Golgi's defenses from being overrun by microbes. At the age of eighty-two, he died in his own bed, on January 21, 1926.

Cajal, too, continued working into old age. After he retired in 1922 from the University of Madrid at seventy, he continued to work in his laboratory and to write. The Spanish government built the Instituto Cajal, far more modern and elegant than the third-floor room in a building next to the Anthropological Museum where he had done most of his work. In 1930, his wife, Silvería, died. Cajal's wife was educated only at the village school, and he shows us little of their intimate life or his feelings when she died. It is clear from what he did write about her that she was his partner in a traditional Spanish household and that he loved her deeply. Moreover, he understood her role in making his life in science possible. He now took his solitary walks in Buen Retiro, the great central park in Madrid, more slowly, and when he returned home the house was empty. For many years, Cajal had taken the same route through this park, past a fountain, on his way to a cafe or his lab. In his last years, he had to change his route to avoid being embarrassed at passing that same fountain, where his statue had been erected. He could sometimes be found sitting in the nearby cafes with friends, and though he no longer sat at the bench with them, his students still came to visit.

All the students who came under Cajal's tutelage heard the
same advice: "It is not sufficient to examine; it is also necessary to observe and reflect: we should infuse the things we
observe with the intensity of our emotions and with a deep
sense of affinity." Two of those students, the histologist Pio
del Rio Hortega (whose views about glial cells so differed
from Cajal's that he left his teacher's lab) and the great neurosurgeon Wilder Penfield, visited their master when he was
eighty. They found him propped up in bed on pillows, busy at
a manuscript that was eventually published as *Neuron Theory or Reticular Theory*. As these two of his most famous students entered the room,

> he put down the goose quill with which he had been writ-
> ing and smiled a courteous welcome. The wall adjacent to
> his impatient right hand was spattered with ink. . . . He was
> feeble, and deafness was closing doors to the world
> around him. But within him there burned the boundless
> enthusiasm of the born explorer and his eyes blazed at us
> through shaggy brows as he talked.

Political passions kept on marching Spain relentlessly
toward civil war. Republicans had taken over the government
again in 1931 but were unable to consolidate their power.
Violently opposed by the army, communists, anarchists, and
Franco, the republic wavered, and general strikes intermittently paralyzed the country. In the spring of 1934, Mussolini
formed a secret pact with the Spanish Fascists, and by October, after the president of the Generalitat in Barcelona pro-

claimed a Catalan Republic, the government declared martial law.

Cajal was still able to discuss the growing political unrest
with his students, children, and grandchildren until a few
hours before the end. He died at home on October 17, 1934.
Ironically, for it would have horrified Cajal, that was the same
day the death penalty was reintroduced in Spain for citizens
accused of "social crimes."

The simplicity of his funeral was a fitting conclusion to his
life. Hundreds of Spaniards from every part of society, including a large representation by trade unions, attended when he
was buried beside his wife in the Necropolis of Madrid.

Cajal chose perfectly when he found the metaphor of
twins conjoined at the shoulders to define his life's entangle-
ment with Golgi. They were forever intellectually aligned and
personally alienated, and all the subsequent discoveries in
neurobiology rest securely on the joined shoulders of these
two nineteenth-century men. Both made heroic discoveries,
and which is the greater will always be argued by their partisans. But it was certainly Cajal who, with a combination of
artistry and passion, first saw into microscopic forms, understood their meaning, and wrested from those images the real
nature of the neuron and the synapse.

Modern neuroscience is now often a collaboration
between teams of researchers who use powerful electronic
technology to discover parts of scientific mysteries. Cajal
worked alone at a simple light microscope and told his own
students that "forming the reckless desire to devote myself to

the religion of the laboratory. . . . I was usually doing things that didn't work, was lost, and more than once hopelessly discouraged. . . . [but] fate doesn't always smile on the rich; from time to time it brings joy to the homes of the lowly." His life was joyous; had he seen nothing more than the synapse, he would have seen enough.

But he did see more. Cajal had what today would be thought of as counterintuitive advice for his students (whom he chose from among the "headstrong") about how to be successful investigators. Scientists should be adventuresome people, he said, restless and imaginative. They should be "generous souls—poets at times, but always romantics—and they have two essential qualities. They scorn material gain and high academic rank, and their noble minds are captivated by lofty ideals."

This is Cajal's own epitaph.

I Have Grown Slow and Palsied

STANDING UP FROM the bench of pure science and walking into neurology, neurosurgery, or psychiatry clinics doesn't take one far from neurons. Once they accepted the neuron theory, clinicians soon began to find illnesses that are the expression of diseased synapses.

By the early 1970s, sophisticated neurochemistry and electron microscopy had evolved enough so that the structure and function of neurons, including the basics of the chemical process by which cells dispatch a message across a gap to their neighbor, became a major research effort. This basic research had immediate clinical applications. One of the earliest of the synaptic diseases to be treated chemically was a debilitating abnormality of the motor endplate (where nerves stimulate muscles) called myasthenia gravis.

In myasthenia, acetylcholine (the same chemical that Loewi found slowed the second toad's heart) is released properly at the presynaptic membrane and crosses the synaptic space as it should, but when it arrives on the other shore the muscle is immune to its signal. The more a myasthenic patient demands of the muscle, the weaker it becomes. This failure occurs in myasthenia because these patients form antibodies that obstruct postsynaptic stimulation by acetylcholine. Similarly, botulinum toxin, that product of careless food preservation, produces related symptoms, though often more abrupt and severe, by preventing the release of acetylcholine on the presynaptic side of the synapse. Here the chemical simply doesn't get into the synaptic cleft. Because the toxin is long acting and eating contaminated food affects the muscles of respiration, botulism may be fatal. On a happier note, injecting tiny doses of botulinum toxin (botox) into specific muscles can now make them relax enough so that wrinkles temporarily vanish and muscle twitching halts!

More recently, psychiatrists have recognized conditions, including some types of depression and compulsion, as occurring when brain stores of neurochemicals such as serotonin become too low. Increasing serotonin concentrations in the synaptic cleft by preventing its reuptake into presynaptic storage vesicles can make people feel better.

Even electrical synapses get their own illnesses. Investigators have found, for example, that X-linked Charcot-Marie-Tooth disease, a disease of peripheral nerves usually characterized by leg weakness, is caused by a mutation on the X chromosome that affects gap junctions. Although various tangles, proteins, filaments, and genes are suspected to play a role, Alzheimer's disease itself may be primarily a failure of the synapse. Memory is ultimately dependent on the plasticity of neurons, their arrangements, and their histories of activation. When the synapses that make new memories fail, for whatever reason, the clinical expression of that failure is dementia.

One of the gravest and best understood of these synaptic illnesses is Parkinson's disease. This condition has been recognizable clinically for nearly two hundred years and requires few imaging studies or lab tests to diagnose. Patients with Parkinsonism are often rigid, usually move and speak very slowly, and their unanimated faces camouflage what their lives had been before they became sick. Intermittently, though, and for no apparent reason, they may also be roused to a hurrying (or festination) of both speech and gait. It was this odd combination of both slowness and tremor, occasionally interrupted by hurry, that caused the great nineteenth-century French neurologist Jean Charcot to err when he called Parkinson's disease a neurosis. This is the same illness so tenderly described by Oliver Sacks in his book *Awakenings*. Those extreme cases included patients with postencephalitic syndromes that developed after the first reported epidemic of the sleeping sickness described in 1916. Fifty years later, Sacks discovered about eighty such patients still institutionalized in a New York hospital for the chronically ill. Most of them had spent years locked on these wards, alive but not really awake or engaged in the world. In his book, Sacks shows us how twenty of them awakened after treatment and describes the spectrum of responses to the great change in their awareness and mobility.

Because there are several surgical treatments for this illness, neurosurgeons often care for patients with Parkinson's disease. When I initially met him as a patient, the physics professor didn't seem like the sort of person who would have been an important developer of the first atomic weapons. Past eighty in 1985, disheveled and withdrawn, he had spent the latter part of his academic career trying to construct a neutrino collector at the bottom of a mine shaft. It isn't clear how many neutrinos he gathered.

Long before, when Robert Oppenheimer began to recruit brilliant young quantum physicists from around the country to the Manhattan Project, the professor had been one of the chosen, lured away from university to help build the "trigger" for the first atomic bombs. He and his colleagues proved themselves capable. The bomb and its triggering device worked perfectly.

After talking to him for ten minutes, I realized that my patient had suffered as much for his involvement in this work as had Prometheus for having stolen fire. Like many of the physicists who worked at Los Alamos (all, perhaps, with the exception of Edward Teller), the professor had come to agree with Oppenheimer, who by the nineteen fifties was publicly denouncing the nuclear age. Earlier, perhaps even at Los Alamos, he had begun to regret his own role in creating it. Oppenheimer's quote on the day of the first nuclear test is well known, but Vannevar Bush recalled that, two nights before the Trinity explosion in 1945, the Manhattan Project director also recited, perhaps to comfort himself, another passage from the *Bhagavad-Gita*:

In battle, in forest, at the precipice of the mountains,
On the dark great sea, in the midst of javelins and arrows,
In sleep, in confusion, in the depths of shame,
The good deeds a man has done before defend him.

Nearly fifty years later, the unhappy professor's regret persisted in a search for enough good deeds done before the Manhattan Project to vindicate him.

But now the physical difficulty of Parkinsonism dominated his life. For reasons unknown, neurons of his substantia nigra had slowly stopped producing dopamine. These areas of the brain (there are two, located symmetrically in the midbrain, one on each side) are black stripes of neurons sandwiched between the pyramidal tracts and the red nuclei. In Parkinson's disease, this dramatic little landscape is the field of a bitter, losing battle because the synapses don't work properly.

The pyramidal tracts connect cells in the cerebral cortex with muscles on the opposite side of the body, while the red nuclei (also bilateral) are two of several neuronal substations deep in the brain that regulate the final expression of movement. Perfect harmony is required between these colorful collections of cells as they regulate each other up and down their intricate pathways to produce smooth voluntary movement of muscles. Dopamine, the key neurotransmitter manufactured in the cells of the substantia nigra, is the conductor orchestrating this intricate process.

After neurons of the substantia nigra produce it, dopamine rides down the axons and is released at synapses in the striatum, another of the deep-brain collections of cells that modulate muscles. The influence of dopamine on the striatum is inhibitory; that is, it lessens the influence of the always eager striatal cells. In normal people, this ballet of cellular authority regulates the activities of complementary muscle groups, resulting in smooth movements: The muscles relax or contract cooperatively as they produce coordinated movements of body parts. But when dopamine is deficient this inhibitory effect is diminished, resulting in the condition first described in 1817 by James Parkinson in his now-famous book, *An Essay on the Shaking Palsy.*

The essential failure in Parkinson's disease is the release of too little dopamine at the synapse. Attempting their delicate conversations with less than they need, the axon of one neuron and the dendritic receptor of its neighbor fail in their colloquy. This failure—too little transmitter and an unstimulated receptor—results in the same kind of dead end as occurs between acquaintances searching for some conversation at a party both of them are trying to leave. Fortunately for my patient the physics professor and others like him (women are affected as often as men), in the 1950s the Swedish neuroscientist Arvid Carlsson discovered the biochemical substance L-dopa, which is metabolized into dopamine. While dopamine itself does not cross the formidable roadblock of the blood-brain barrier, L-dopa slips through easily, restoring brain levels of dopamine and thereby easing symptoms of the disease. With L-dopa replacement therapy, the physicist could function well enough to continue his search for neutrinos.

While doctors had a certain amount of success in treating patients with Parkinsonism more effectively after L-dopa became readily available, it was not a cure-all. Since the drug only affects some of the symptoms, neurosurgeons continue to operate on selected patients with severe tremor or rigidity. The original surgery to treat these symptoms employed the same kind of technology used in the animal experiments at the Chicago lab where I had worked as a student. Using trigonometry to locate the correct group of cells, we placed an electrode deep into the brain and passed a brief current through it. This stereotaxic lesion in the thalamus—a large, busy thoroughfare just above the brain stem with millions of neurons both coming and going—adjusted the circuit. By killing some of the cells in one of the thalamic nuclei, the operation often smoothed out movements on the opposite body side and stilled the tremor. If the patient had extreme symptoms on both sides of the body, the operation had to be done on both sides of the brain, increasing the risks.

More recently, the surgical targets have changed slightly, and often the cells are now stimulated with implanted electrical generators rather than killed, but the idea of rebalancing the neural mechanisms for movement is the same. In essence these procedures create a new harmony among the several motor systems, reducing the excitatory influences of a key relay station in the thalamus to compensate for the death of some of the dopamine-producing cells downstream in the substantia nigra.

In some ways, the entire field of neurosurgery became possible because of the neuron theory. While some bold surgeons had performed simple operations to remove abscesses or blood clots from around the cranium or spine, few before Victor Horsley were brave or foolish enough to actually do much to the brain or spinal-cord tissues themselves. After all, if the nervous system was all connected as a network, operating on one place might have unforeseen consequences at a distant site. Surgeons worried that perhaps this network might work like a string of Christmas-tree lights: A blowout in one bulb might disable the entire string.

The theory that individual units turn on and off to transmit information established a new paradigm, and soon after his trip to London, Harvey Cushing began to operate on patients in Baltimore and to train the first generation of American neurosurgeons. By the end of World War II, outcomes improved dramatically, and neurosurgical procedures had become routine. But the reign of neurosurgery as the king of specialties has been brief. Because of rapid technological advances in applied therapeutics, imaging, and nonsurgical therapies such as focused-beam irradiation and endo-vascular treatment for cerebro-vascular diseases, only one hundred years after Cushing such operations are less and less required.

The discovery that made the neuron theory law and eventually allowed the remarkable treatment of Parkinson's disease and other diseases of the synapse, as well as surgery on the brain, was Cajal's positive identification of the "gap." Exposing this gap permitted a growing understanding of how neurons transmit information electrically and chemically, and opened the way for modern neurobiology.

The work of the "conjoined twins," a dogged and reserved boy from the Italian Alps and a passionate Spanish boy from the Pyrenees, changed the way we see our own brains. Freud, who began his career as an anatomist just when the neuron controversy was at its height, spoke as a scientist when he claimed that what, in one moment, is praised as the highest wisdom is soon rejected and replaced by "the latest error [that] is then described as the truth." He had become a psychiatrist by the time he wrote, "The communal life of human beings [has], therefore, a two-fold foundation: the compulsion to work, which was created by external necessity, and the power of love." Cajal, the scientist and artist, would have agreed with both these observations. Even when he was confined to bed and preparing to die, he optimistically loved his students, family, and friends and continued to work. "I neither will nor should cease my efforts," he wrote. "I have besides the unavoidable duty of directing my pupils, inspiring them with unquenchable confidence in their own powers and robust faith in indefinite progress. Science, like life, grows ever, renewing itself continually. . . . It is a great stimulus to the young to know that the mine is inexhaustible. "

Notes

Unless otherwise noted, all the direct quotes attributed to Cajal are taken from Santiago Ramón y Cajal, *Recollections of My Life* (Cambridge, Mass.: MIT Press, 1966).

Preface

18 "For all those": Cajal 604.

Chapter 1: A New Way to See

29 "Instead of elaborating": Cajal (1999) 1.
31 "We Germans fear": Massie 90.
34 "Man is to himself": Pascal 31.
35 "wondrous small creatures": Marshall and Magoun 127.

Chapter 2: The Crazy Navarran

41 "Aside from the usual cases": Flaubert 81.
42 "Oh, the heroic rustics": Cajal 11.
43 "recesses and windings": Ibid., 15.
43 "most disgusting affectation": Cannon 13.
43 "little gentlemen": Cajal 27.
43 "certain morbid pleasure": Ibid., 16.
44 "My appearance in the public square": Ibid., 26.
44 "In the bottom of every": Ibid., 30.
45 "absolutely essential preparation": Ibid., 29.
45 "madness over art": Ibid., 53.
45 "the active to the contemplative ones": Ibid., 39.
46 "the Aristarch": Ibid., 40.

46 "*ex cathedra*": Ibid., 41.
47 "mature deliberation": Ibid., 70.
48 "who had already complaints": Ibid., 71.
48 "an overflowing flora and fauna": Ibid., 72.
49 "positively stupefied me": Ibid., 140.
49 "We were incorrigible": Ibid., 73.
50 "timorous child": Ibid., 117.
50 "the crazy Navarran": Ibid., 97.
51 "within a short time you will rise": Ibid., 108.
51 "not had any master": Ibid., 159.
51 "a manufacture of his own": Ibid.
52 "My pencil": Ibid., 169.

Chapter 3: A Laboratory in the Kitchen

53 "the so-called *draft of Castelar* ": Ibid., 195.
54 "heroic doses of quinine": Ibid., 218.
54 "confronted the offending officers": Cannon 73.
55 "Who are you to disobey me?" Cajal 231.
55 "catastrophe": Ibid., 226.
55 "the tremendous blunders": Ibid.
55 "who absconded to the United States": Ibid., 221.
56 "the amazing spectacle": Ibid., 250.
56 "are gifts from God": Cajal (1999) XIV.
57 "As with the lover": Ibid., 112.
58 "system of treatment": Cajal 266.
59 "great green eyes framed": Ibid., 270.
59 "spiritual director": Ibid., 272.
60 "pretty weak": ibid., 275.
64 "I am a fool!": Ibid., 327.

Chapter 4: Cells, Fibers, and Networks

67 "I wish to know what the primitive": Marshall and Magoun 126.
69 "the entire nervous system": Ibid., 128.
69 "the fibers originated": Ibid.
74 "It appeared as if nature": Mazzarello 41.
75 "a cell-state": Garrison 570.
76 "*structure and arrangement*": Virchow 228.
76 "is a very erroneous one": Ibid., 229.

76 "For they fancied they saw": Ibid.
78 "amorphous matrix": Mazzarello 46.
79 "degradation of the fibrillary substance": Ibid., 47.
79 "the elegant shape": Ibid.
79 "The interstitial stroma": Ibid.

Chapter 5: The Black Reaction

81 "began to bully his son": Mazzarello 55.
81 "honorable and stable position": Ibid.
83 "it is clear that these conflicting": Ibid., 64.
84 "What a fantastic sight!": Ibid., 1.
85 "Delighted that I have found": Ibid., 63.
86 "With great emotion": Ibid., 65.
86 "collaterals of the cylindraxis": Ibid., 119.
88 "represented a great theoretical": Ibid., 81.
93 "feelings never before experienced": Ibid., 104.
94 "partly determined by chance": Ibid., 80.
96 "for a man dedicated": Cajal 316.
97 "new truth": Ibid., 322.

Chapter 6: The Trench of Science

99 "the devil of pride": Ibid., 328.
99 "did not move a pawn": Ibid.
100 "strategic subterfuges": Ibid., 323.
102 "with her self-denial": Berciano 119.
102 "pursuit of chimeras": Cajal (1999) 1.
103 "great men are at times geniuses": Ibid., 9.
106 "claw-like arborizations": Cajal 330.
107 "Every part of the cortex": Marshall and Magoun 132.
109 "the new cartography of mind": Harrington 210.
111 "to work, no longer merely": Cajal 325.
112 "the trench of science": Ibid., 321.

Chapter 7: Climbing Fibers and Basket Endings

115 "the so-called protoplasmic processes": Mazzarello 119.
115 "How much of what we see": Laqueur 20.
115 "a system of ramifications": Mazzarello 89.
117 "there is no way of knowing": Ibid., 211.

117 "were not even cited": Ibid., 204.
120 "to find in the life of my uncle": Ibid., 105.
121 "robust conductors": Cajal 332.
121 "This fortunate discovery": Ibid.
122 "the intellectual lucidity": Cajal 278.
122 "have been found as a result of chance": Cajal (1999) 66.
123 "terminate with small bulges": Mazzarello 213.
123 "receive and propagate": Cajal 322.
125 "gathered by certain bipolar cells": Cajal (1990) 106.
127 "perfectly irreconcilable schemes": Ibid., 346.
127 *"law of pericellular contact"*: Ibid., 342.
127 "applied intimately to the contours": Ibid., 346.

Chapter 8: Sincere Congratulations Burst Forth

134 "a rascally young fop": Massie 92.
138 "Finally, the prejudice against": Cajal 356.
139 "Cajal's first studies": Mazzarello 196.
139 "I have discovered": Berciano 119.
140 "a state of savagery": Cajal 360.
141 "my preparations are so clear": Mazzarello 195.
145 "I cannot believe": Ibid., 204.
145 "impossible and superfluous": Ibid., 212.
146 "I admire the work of Golgi": Ibid., 206.

Chapter 9: The Prize

149 "A theory is not an unemotional thing": Cannon 122.
149 "the retina has always": Cajal 392.
150 *"the theory of dynamic polarization"*: Ibid., 389.
154 "Who could make a lady": Ibid., 486.
154 "madmen discharging rifles into the air": Ibid., 488.
157 "had his patient under ether": Fulton 163.
157 "yields a better adjectival form": Marshall and Magoun 144.
158 "intense anthropomorphism": Cannon xiii.
158 "he treated the microscopic": Ibid.
160 *"Carolinische Institut verliehen Sie Nobelpreis"*: Cajal 545.
160 "this nomination has previously been made": Corral 43.
160 "How could I justify": Cajal 549.
161 "satisfaction at finding myself": Mazzarello 310.

161 "There was no better occasion": Ibid., 312.
162 "Every well-bred person must": Cajal 547.
163 "prudish daughters of the north": Cajal (2001) 6.
164 "always been opposed to the neuron theory": Golgi 189.
164 "this doctrine is generally": Ibid.
164 "the idea of the protoplasmic process": Ibid., 205.
165 "looked at the speaker with stupefaction": Mazzarello 316.
166 "there is a fear": Morris 582.
167 "In strict confidence": Zinn 290
167 "*necessary* war": Morris, 593.
167 "Is it not the acme of irony and humor": Cajal 550.
167 "the will to shine": Brinton 27.
167 "*homo sapiens* only philosophizes": Cajal (2001) 117.

Chapter 10: Higher Magnification

170 "synaptic substance": Marshall and Magoun 165.
176 "But by the Triassic times": Romer 511.
176 "That the brain has ten billion neurons": Marshall and Magoun 156.
181 "He showed, for instance, that each nerve path": Stevens 62.
182 "constitute a small but respectable minority": Bennett 27.
183 "They can do many things": Ibid., 16.
183 "law is nothing more or less": Menand 347.
183 "he applied to his own special field": Ibid.

Chapter 11. Recollections of Two Lives

184 "What a cruel irony of fate": Cajal 553.
185 "strong-willed, smooth talking": Mazzarello 251.
186 "was tantamount to accusing Golgi": Ibid., 326
186 "beautiful preparations": Ibid.
186 "specific function that is thought to belong": Ibid.
186 "the fact is not less true": Ibid., 327.
186 "glorious events in which we have had": Ibid., 330.
187 "proud to belong to the ruling class": Ibid.
188 "romantic, argumentative, and aggressive": Ibid., 347.
188 "it was a stinging and brilliant story": Ibid.
189 "the morbid desire to assert": Cajal 563.
189 "to wrest new truths from it": Ibid., 564.

189 "against other men": Ibid.
189 "dislodges some deeply rooted error": Ibid.
189 "the incomparable observer": Fabre 25.
190 "dear insects": Ibid., xiv.
190 "not only instinct": Cajal 579.
190 "internal impulses": Ibid., 589.
190 "Why call up pains": Ibid., 581.
191 "I owe to photography": Craigie and Gibson 199.
191 "adhering to receipts and formulae": Cajal 581.
191 "Like every old man": Ibid., 604.
192 "bloody crises of civilization": Ibid., 583.
194 "It is not sufficient to examine": Cajal (1999) 112.
194 "he put down the goose quill": Cajal (1954) x.
195 "social crimes": Paz 62.
195 "forming the reckless desire": Cajal (1999) xiii.
196 "generous souls—poets at times": Ibid., 142.

Epilogue

201 "In battle, in forest": Rhodes 663.
205 "the latest error": Freud (1964) 171.
205 "The communal life of human beings": Freud (1961) 55.
205 "I neither will nor should": Cajal 604.
205 "I have besides the unavoidable duty": Ibid.

Glossary

Action potential. To "turn on" the cell. The electrical *depolarization* (discharge) of a neuronal membrane resulting in synaptic transmission, first characterized by Émil du Bois-Reymond in 1848.

Axon. The single fiber of a neuron that conducts an electrical impulse away from the cell body and toward a synapse on another cell's dendrite. Also sometimes called *nerve fiber* and *cylindraxis* by early histologists. The term *axon* was proposed by Kölliker in 1896.

Brain stem. The part of the brain composed of fiber tracts and cell bodies connecting the cerebral cortex and cerebellum above with the spinal cord below. The brain stem has three divisions from bottom to top: medulla, pons, and midbrain. Sensory information entering the brain itself and motor messages leaving it traverse the brain stem. The collections of neurons in the brain stem are the origins of cranial nerves, which have specialized functions related to vision, eye movements, hearing, sensation on the face, swallowing, etc.

Boutons. The bare axonal endings that terminate in a synapse on the next cell.

Cerebellum. This word is the Latin diminutive of *cerebrum*, therefore, little brain. It is the part of the brain at the back of the head separated from the cerebral cortex by a thick membrane above and the brain stem in front (ventral). The cerebellum is functionally related to coordination and balance.

Contact. A term used by Cajal and others meant to imply transmission of electrical impulses between individual *neurons* as opposed to a continuous *network* of fibers.

Cylindraxis. An early name for *axons* coined by Joseph Rosenthal, a student of Purkinje's. Also sometimes called axis cylinder process. The term was later employed by both Golgi and Cajal to describe what are now called *axons* and bare axonal endings now called *boutons*.

Dendrite. The fibers that arise from the bodies of neurons and conduct electrical impulses into the cell. The surface of the dendrite is covered by knobby extensions called *spines,* where *synapses* occur. Dendrites were sometimes called *protoplasmic processes* by early histologists. The term *dendrite* was first proposed by Wilhelm His, Sr., in 1889.

Depolarization. A change in the normal *resting potential* of a cell that results from movement of ions across the membrane.

Electrical gradient. The voltage maintained across a neuronal membrane. This potential electrical difference is caused both by passive diffusion and by the active transport of ions into and out of the cell. Active transport requires energy in the form of ATP to pump ions against the gradient, in essence storing dissimilar metal ions (mainly sodium and potassium) and making a battery.

Gap junction. An area between certain kinds of neurons where the axon and dendrite are so close that information passed between them is electrical, not chemical, as is the more common case.

Myelin. A laminated wrapping around nerve fibers composed mostly of fat and protein; an insulation on axons. The insulation is interrupted at regular intervals by narrowings called *nodes of Ranvier,* and the rate of nerve conduction is increased by skipping from node to node. Schwann cells produce myelin around peripheral nerves, while glial cells form the myelin around axons in the brain and spinal cord. Because of the high fat content, myelin looks white—thus, white matter.

Neuroglia. Supporting cells of the central nervous system that are not neurons. Some glial cells form myelin around axons in the brain and spinal cord, and some contribute to the repair of injured cells. These cells also play a role in forming the blood-brain barrier and help exclude certain chemicals from freely entering the brain from blood.

Neuron. The kind of cell located in the brain and spinal cord consisting of a cell body, dendrites, and a single axon. This is the fundamental unit for transmitting information in the nervous system. Dense clusters of neurons without a myelin coating look gray, leading to the term *gray matter.* Waldeyer did not name the neuron until 1891, but I have occasionally used it in the text when discussing brain cells prior to that date.

Postsynaptic membrane. The receiving surface membrane on cell bodies or dendrites that is stimulated across the synapse.

Presynaptic membrane. The generating surface membrane on axons that ends as a synaptic *bouton* (or knob) and stimulates across the synapse. This was the last part of the structure nineteenth-century histologists saw as the *cylindraxis*.

Resting potential. The value of an *electrical gradient* across a neuronal cell membrane when it is "turned off." At rest, the inside of the cell is negative because of the way concentrations of ions are maintained inside and outside the cell. (See *Electrical gradient,* above.) *Depolarization* of the cell membrane alters this gradient by changing ion concentrations and, therefore, produces an *action potential* down the axon ending in the synapse.

Reticulum (synonymous with network). The theory that the entire nervous system is anatomically connected, especially that each neuron is connected directly to every other neuron.

Rods and cones. The light-receiving cells of the retina in the eye.

Synapse. A cleft or gap bordered by the membrane of an axon and on the other receiving side by the membrane of a dendrite or cell body. The typical junction is chemical. These chemicals either stimulate or inhibit the postsynaptic membrane. The sum total of these stimulatory and inhibitory inputs determines whether or not the next cell fires its own *action potential.* The size of the synaptic cleft varies but is about 200 to 300 angstroms (200 to 300 millionths of a centimeter).

Thalamus. An enormous collection of cell bodies deep in the cerebral cortex of the brain that modulates and connects both motor and sensory functions. The thalamus is further subdivided into nuclei based on cellular anatomy and function.

Bibliography

Anderson, C. G., and B. Anderson. 1993. "Koëlliker on Cajal." *International Journal of Neuroscience* 70: 181–92.

Bennett, M. R. 1999. "The Early History of the Synapse: From Plato to Sherrington." *Brain Research Bulletin* 50(2): 95-118.

Bennett, M. V. L. 2000. "Electrical Synapses: A Personal Perspective (or History)." *Brain Research Reviews* 32: 16-28.

Berciano, J., M. Lafarga, and M. T. Berciano. 2001. "Santiago Ramón y Cajal." *Neurología* 16(3): 118–21.

Brinton, C. *A History of Western Morals.* New York: Harcourt, Brace and Company, 1959.

Cannon, D. *Explorer of the Human Brain: The Life of Santiago Ramón y Cajal.* New York: Henry Schuman, 1949.

Castiglioni, A. *A History of Medicine.* New York: Knopf, 1958.

Clark, E., and C. D. O'Malley. *The Human Brain and Spinal Cord.* Berkeley: University of California Press, 1968.

Corral, I. C., C. C. Corral, and A. C. Castanedo. 1998. "Cajal's Views on the Nobel Prize for Physiology and Medicine (October 1904)." *Journal of the History of the Neurosciences* 7(1): 43–49.

Craigie, E. H., and W. C. Gibson. *The World of Ramón y Cajal.* Springfield, Ill.: Thomas, 1968.

Eccles, J. C. *The Understanding of the Brain.* New York: McGraw-Hill, 1977.

Eiseley, L. *Darwin's Century: Evolution and the Men Who Discovered It.* New York: Doubleday, 1958.

Fabre, J. H. *The Insect World of J. Henri Fabre.* Boston: Beacon Press, 1991.

Falk, M. M. 2000. "Biosynthesis and Structural Composition of Gap Junction Intercellular Membrane Channels." *European Journal of Cell Biology* 79: 564–74.

Flaubert, G. *Madame Bovary.* Translated by Mildred Marmur. New York: Doubleday, 1997.

Freud, S. *Civilization and Its Discontents.* Translated and edited by J. Strachey. New York: W. W. Norton, 1961.

———. *The Standard Edition of the Complete Psychological Works of Sigmund Freud.* Volume 22. Translated and edited by J. Strachey. London: Hogarth Press, 1964.

Fulton, J. F. *Harvey Cushing: A Biography.* Springfield, Ill.: Thomas, 1946.

Garfield, S. *Mauve.* New York: W. W. Norton, 2001.

Garrison, F. H. *An Introduction to the History of Medicine.* Philadelphia: Saunders, 1929.

Gilman, S. 1997. "Alzheimer's Disease." *Perspectives in Biology and Medicine* 40(2): 230–45.

Golgi, C. 1906. "The Neuron Doctrine—Theory and Facts." In *Nobel Lectures, Physiology or Medicine 1901–1921.* New York: Elsevier, 1967.

Grey, E. C. 1969. "Electron Microscopy of Excitatory and Inhibitory Synapses: A Brief Review." In K. Akert and P. Waser (eds.), *Progress in Brain Research, Mechanisms of Synaptic Transmission* 31: 141–55.

Haymaker, W. *The Founders of Neurology.* Springfield, Ill.: Thomas, 1953.

Harrington, A. "Beyond Phrenology: Localization Theory in the Modern Era." In *The Enchanted Loom,* edited by P. Corsi. Oxford: Oxford University Press, 1991.

Jones, E. G. 1999. "Golgi, Cajal and the Neuron Doctrine." *Journal of the History of the Neurosciences* 8(2): 170–78.

Lain Entralgo, P. 1978. "Ramón y Cajal 1852–1934." Madrid: Expediénts Administrativos de Grandes Españoles.

Laqueur, T. 2003. "Emil Mayer VIII." *Three Penny Review,* Winter: 20.

Marshall, L. H., and H. W. Magoun. *Discoveries in the Human Brain.* Totowa, N.J.: Humana Press, 1998.

Massie, R. K. *Dreadnought.* New York: Ballantine, 1991.

Mazzarello, P. *The Hidden Structure: A Scientific Biography of Camillo Golgi.* London: Oxford University Press, 1999.

Menand, L. *The Metaphysical Club.* New York: Farrar, Straus and Giroux, 2001.

Meyer, A. C. *Historical Aspects of Cerebral Anatomy.* London: Oxford University Press, 1971.

Meynert, T. 1885. "Psychiatry: A Clinical Treatise on Diseases of the Fore-Brain Based Upon a Study of its Structure, Function and Nutrition." Translated by B. Sachs. New York: Putman, 1968.

Mitchner, J. A. *Iberia.* New York: Random House, 1968.

Morris, E. *The Rise of Theodore Roosevelt.* New York: Modern Library, 2001.

Pascal, B. *Thoughts and Minor Works.* The Harvard Classics, edited by Charles Eliot. New York: P. F. Collier & Son, 1910.

Paz, A. *The Spanish Civil War.* Paris: Hazan, 1997.

Poritsky, R. 1969. "Two and Three Dimensional Ultrastructure of Boutons and Glial Cells on the Motoneuronal Surface in the Cat Spinal Cord. *Journal of Comparative Neurology* 135: 423–52.

Ramón y Cajal, S. *Advice for a Young Investigator.* Cambridge, Mass.: MIT Press, 1999.

———. 1892. "El neuvo concepto de la histologia de los centros nerviosos. III—Corteza gris del cerebro." *Scientific Medical Review of Barcelona* 18: 457–76. Translated by D. A. Rottenberg in D. A. Rottenberg and F. H. Hoffberg, eds., *Neurological Classics in Modern Translation.* New York: Hafner, 1977.

———. *Recollections of My Life.* Cambridge, Mass.: MIT Press, 1966.

———. *Recuerdos de Mi Vida.* Madrid: Juan Pueyo, 1923.

———. *Neuron Theory or Reticular Theory?* Translated by M. U. Purkiss and C. A. Fox. Madrid: Instituto Ramón y Cajal, 1954.

———. *New Ideas on the Structure of the Nervous System in Man and Vertebrates.* Cambridge, Mass.: MIT Press, 1990.

———. *Vacation Stories: Five Science Fiction Tales.* Urbana: University of Illinois Press, 2001.

Rhodes, R. *The Making of the Atomic Bomb.* New York: Simon and Shuster, 1986.

Robinson, J. D. *Mechanisms of Synaptic Transmission.* Oxford: Oxford University Press, 2001.

Romer, A. S. *The Vertebrate Body.* Philadelphia: Saunders, 1955.

Sacks, O. *Awakenings.* New York: Vintage Books, 1999.

Shepherd, G. M. *Foundations of the Neuron Doctrine.* New York: Oxford University Press, 1991.

Singer, S., and E. A. Underwood. *A Short History of Medicine.* New York: Oxford University Press, 1962.

Smith, C. U. M. 1996. "Sherrington's Legacy: Evolution of the Synapse Concept, 1890s–1990s." *Journal of the History of the Neurosciences* 5(1): 43–55.

Soriano, V. 1981. "History of Neural Transmission." *International Journal of Neurology* 15(1–2): 132–52.

Stevens, L. *Explorers of the Brain.* New York: Knopf, 1971.

Szentágothai, J. "What the 'Reazione Nera' Has Given to Us." In *Golgi Centennial Symposium,* edited by Maurizio Santini, 1–12. New York: Raven Press, 1975.

Toynbee, A. *Mankind and Mother Earth.* New York: Oxford University Press, 1976.

Virchow, R. *Cellular Pathology.* Birmingham, Ala.: The Classics of Medicine Library, 1978.

Whitney, C. *The Discovery of Our Galaxy.* New York: Knopf, 1971.

Williams, H. *Don Quixote of the Microscope.* London: Jonathan Cape, 1954.

Zinn, H. *A People's History of the United States.* New York: Harper Colophon, 1980.

Index

Page numbers in *italics* refer to illustrations.